A TQM Approach
to Achieving
Manufacturing Excellence

A TQM Approach
to Achieving
Manufacturing Excellence

A. Richard Shores

QUALITY PRESS
American Society for Quality Control
Milwaukee, Wisconsin

QUALITY RESOURCES
A Division of The Kraus Organization Limited
White Plains, New York

Copyright © 1990 ASQC Quality Press

Printed in the United States of America

94 93 92 91 90   10 9 8 7 6 5 4 3 2 1

ASQC Quality Press
310 West Wisconsin Avenue
Milwaukee, Wisconsin 53203

Quality Resources
A Division of The Kraus Organization Limited
One Water Street, White Plains, New York 10601

Library of Congress Cataloging-in-Publication Data
Shores, A. Richard, 1942–
    A TQM approach to achieving manufacturing excellence / A. Richard
Shores.
        p.   cm.
    Includes bibliographical references and index.
    ISBN 0-527-91632-3
    1. Production management—Quality control.   I. Title.
TS156.S49   1990
658.5′62—dc20   90-8865
                                                                  CIP

ISBN 0-87389-109-0 (ASQC Quality Press)
ISBN 0-527-91632-3 (Quality Resources)

Contents

Figures

Tables

Preface

Every year there are thousands of articles printed in various magazines and books that provide new or different insights into successfully managing business. Multiplied over the years, there must be hundreds of thousands of "ideas" for achieving success. The pure volume of information can overwhelm managers who must sort through the plethora of literature and make strategic business decisions. In some cases, the integration of this information has broken down, and chaos has been the result. In other situations, businesses have managed their learning well and have established a position of excellence among their peers. These businesses have become the "best of the best" in their industries. They serve as an example for others to emulate.

During the early years of American industry, management contributions came from many businesses. In the first years of the automotive industry, for example, the Ford Motor Company was distinguished by unparalleled gains in productivity through mass-production techniques. John H. Patterson led National Cash Register Company into the forefront of industry with progressive ideas about human dignity and employee rights. IBM proved the value of quality and customer service. Others, like E. I. du Pont, set the pace for technology and innovation.

In later years, the "best of the best" management practices were refined and integrated. The leading companies who led this effort also prevailed in their competitive environments. They became known as the "excellent companies" by Tom Peters (1982) and "The Survivors" by Harry G. Stein (1986). Other authors have similarly dubbed the few elite companies who set the pace for American industry and competitiveness. Managers everywhere now look to these companies as models of corporate excellence.

Since 1980, "new" management ideas have been introduced to American industry. These concepts emerged from Japan under the name of company-wide quality control (CWQC) and total quality control (TQC). Japan's success with these management practices has established new standards for competitive excellence. New models of excellence are now pursued in the names of Toyota, Sony, and Hitachi.

American companies have subsequently begun to add the concepts of TQC into their existing management cultures. This amalgamation is resulting in what many managers now refer to as "total quality management" (TQM). As history has proven, some businesses will be more successful with TQC than others. Those who succeed will become the models of excellence for tomorrow.

In 1987, President Ronald Reagan commissioned a national quality award named after the late Secretary of Commerce, Malcolm Baldrige. This award is intended to recognize companies for competitive excellence. U.S. Secretary of Commerce C. William Verity credits the winners of the Baldrige Award with "showing the way for all of corporate America to improve quality, strength, and market position" (*Quality Progress*, 1989).

In 1988, Motorola, Inc., along with Westinghouse Corporation's Nuclear Fuel Division and Globe Metallurgical Inc., won the first Malcolm Baldrige Award. The three recipients of the award were the best of sixty-six other businesses who had applied, which included forty-five manufacturing companies, twelve small businesses, and nine service companies. In presenting the award at the White House, President Ronald Reagan commented that "They [award winners] realize that customer satisfaction through better quality is the goal. And they know that America's economic strength and future depend more and more upon the quality of its products."

Motorola, Westinghouse Nuclear Fuel Division, and Globe Metallurgical have clearly earned themselves a position of respect among the leading companies of America. The prestige of this award and continued competitive pressure will provide many other businesses the motivation to follow in their footsteps. Other companies in pursuit of this goal will find that the path to success is not an easy objective to reach. They will be confronted with thousands of ideas about how to proceed. They will realize that winning the Baldrige Award or simply becoming the "best among the best" is a multiyear task. They will also learn that winning the award is the secondary benefit—the real benefit is derived from improved competitive performance.

In Motorola's quest to win the Baldrige Award, it pursued a multifaceted program. The company's quality goal is zero defects in everything they do. This includes design, marketing, manufacturing, and service. They also focus on cycle-time reduction in design, marketing, manufacturing, and delivery of products. Motorola's management is committed to top-level leadership and review of their quality efforts. Their nonmanagement employees are involved through a participative management program (PMP). Motorola invested over $170 million in worker-training programs from 1983 to 1987. Consider the magnitude of these strategies when they are applied across all of Motorola, Inc., which is a business of 99,000 employees working at fifty-three major facilities worldwide.

This book is organized as a blueprint for achieving manufacturing excellence. In most respects, it is based on the requirements of the Baldrige Award. But it goes much further by describing strategies to achieve excellence. It is designed to show the complete integration of TQC principles with traditional excellent management practices. The virtue of TQM is self-evident by the parallel thinking that exists between the Baldrige Award and the results of my research. Both were pursued from independent paths. It is my expectation that managers pursuing either competitive excellence, the Baldrige Award, or both, will be satisfied with this book.

Managers reading this book will find an organized approach to pursuing the many aspects of past and present management practices. It is based on the total quality management model that I first introduced in 1988 (Shores, 1988). That book was the result of four years of intensive research and the successful application of total quality management principles while working as a Quality Manager at Hewlett-Packard Company.

A TQM Approach to Achieving Manufacturing Excellence uses an enhanced version of the TQM model from the previous book and applies it to achieving manufacturing excellence. In this application, I have more thoroughly addressed the interrelationships of just-in-time, quality, computer integrated manufacturing, and participative management. My effort has been aided by twenty years experience in manufacturing, research and development, marketing, and quality assurance. I have also, however, found it beneficial to visit other companies and to spend many hours researching and reading almost everything printed and current about the "Factory of the Future."

This research has carried into many fields such as economics, quality, human resource management, computer integration strategy, robotics, total quality control, systems analysis, statistics, and management science. I hope to pass on to you what I have learned from this research: all of these fields are intensely interrelated within an organized system that makes them easy to understand and apply. This is the contribution of *A TQM Approach to Achieving Manufacturing Excellence*.

REFERENCES

Peters, Thomas J., and Waterman, Robert H. 1982. *In Search of Excellence*. New York: Harper and Row.

Quality Progress. 1989. "Reagan Lauds First Baldrige Award Winners." 22(1) (January):525–527.

Shores, A. Richard. 1988. *Survival of the Fittest: Total Quality Control and Management Evolution*. Milwaukee, WI: Quality Press.

Stine, Harry G. 1986. *The Corporate Survivors*. New York: AMACOM.

Acknowledgments

As is usually the case, many people have contributed to bring this book to market. To all of you, I give my special "thank you" for your help. Some people contributed directly by sharing their thoughts and experiences in the Appendixes. As contributing authors, credit is given in the table of contents and in each appendix. Throughout the book, I have tried to be diligent in citing the references where I have benefited from the thinking and writings of others. Again, thank you.

Special thanks go to those who have helped me with the manuscript: my wife, Paula, who has proven to be an excellent reader and editor; my secretary, Ilene Helman, who helped in editing and formatting my final manuscript; and, lastly, my five children, Richard, Karen, Cheryl, Laura, and Autumn. I thank them for their love and support.

Part I

A Total Quality Management Strategy

1

Manufacturing Competitiveness

As we enter the last decade of the twentieth century, U.S. manufacturing is struggling for survival. Increasing shares of our markets (domestic and foreign) are being lost to foreign competitors. The quality of U.S. manufacturing is no longer viewed as "world class." Productivity growth is lagging behind many of our major trading partners like Japan and West Germany. Innovation is lacking, as is creativity and timeliness of new products. In every measure of manufacturing competitiveness, the United States is falling farther behind.

The situation for U.S. manufacturing has deteriorated considerably since the beginning of the twentieth century. Then, U.S. industry was in its infancy and its growth was exhilarating. Industrial dynasties were being created by Henry Ford, John D. Rockefeller, W. C. Durant, and John H. Patterson, to name only a few of the great industrial leaders. The role of production and productivity in the economy was considered fundamental, as Henry Ford wrote: "The foundations of society are the men and means to *grow* things, to *make* things, and to *carry* things. As long as agriculture, manufacture, and transportation survive, the world can survive any economic or social change" (Ford, 1922). Henry Ford was not an economist, but a mechanical genius. Even still, his economic philosophy served society well through most of the twentieth century, generating economic growth, jobs, and prosperity through a strong manufacturing industry.

3

Today, the U.S. economy is in jeopardy. Corporate and personal income tax revenues have not kept pace with federal spending, leading to a record federal deficit of $220 billion in 1986. The foreign trade balance has steadily worsened, reaching a record deficit of over $170 billion in 1987. Manufacturing jobs declined from 20.1 million in 1980 to 17.5 million in 1986; only recently has the low value of the dollar begun to have a reverse effect on manufacturing employment, increasing to about 19 million workers in 1989. Many economists believe that the low growth rate of manufactured products is a large part of today's economic problem.

Statistics from the Department of Commerce show that the real growth rate of the gross national product (GNP) has averaged about 2.7% per year since 1970. The real growth rate for manufacturing from 1970 to 1986 was only slightly above the average at 2.9%. Manufacturing's growth rate was higher than agriculture and mining, but considerably lower than the services sector. Subsequently, the manufacturing share of the GNP has declined from 25% in 1970 to 19% in 1986. Manufacturing's growth rate has been hindered by low growth in basic industrial manufacturing and aided by high growth in high-tech manufacturing. High-tech production has been the only bright star in the manufacturing sector during the last fifteen years.

While manufacturing's growth has been sidetracked, the federal government, corporations, and individuals have continued to spend in social services, communication, banking, utilities, insurance, and health care to create strong growth in the services sector of the economy. Since 1970, the services sector has grown at approximately a 3.5% rate, increasing from 50% of the GNP in 1970 to 60% in 1986. Virtually all of the employment growth in the United States during the last fifteen years has been in the services sector.

During its strong growth, the services sector has invested heavily in high-tech capital goods. The investment is for computers, communications, and medical equipment required to support increasing customer demands for service and a more competitive environment. The heavy investment in high-tech goods *should* drive increases in high-tech manufacturing, which it has done to some extent. Recently, however, increasing amounts of these high-tech goods are coming from foreign manufacturers, contributing further to the trade deficit and the decline of manufacturing.

This trend was noted by Stephen S. Roach in 1986 when he reported that during a two-year period from the fourth quarter of 1982 to the third quarter of 1984, the amount of high-tech imports doubled from $19.7 billion per quarter to $42.7 billion (Roach, 1986). As recently as January 5, 1989, *The New York Times* reported that the American Electronics Association had stated that the United States' share of the worldwide electronics production had dropped from 50.4% in 1984 to 39.7% in 1987. In the same period, Japan's share rose from 21.3% to 27.1%, and Western Europe's share rose from 23.5% to 26.4%. This is a startling statistic, and an indication that the last bastion of U.S. manufacturing superiority is falling.

There has been much political and economic controversy lately about the relationship between the federal deficit and the trade balance. Conventional economic theory says, in summary, that the budget deficit creates a shortage of national savings, which pushes up real interest rates, which attracts foreign investment, which strengthens the dollar, and, therefore, leads to poor international competitiveness and a trade deficit (Hudson, 1987). This theory implies that governments are in full control of the economy. Simple adjustments to spending and/or taxes are all that are required to maintain a balanced budget and a trade surplus. In fact, during the 1988 presidential campaign, about the only economic issues of debate were based on choices of where to cut government spending and/or whether to raise taxes.

COMPETITIVENESS AND PRODUCTIVITY

Peter L. Bernstein is among a growing minority of economists who refute conventional economic theory. Bernstein contends that low productivity relative to worldwide competition leads to the trade deficit. The trade deficit leads to low corporate profits and lost jobs, and then to low tax revenues and low savings. The result is a federal budget deficit, higher interest rates, and the strengthening of the dollar. The high value of the dollar combines with low productivity to compound the international competitiveness problem and the trade deficit (Coy, 1989).

Bernstein's theory appeals to my pragmatic, economic senses because it ties the deficit to the economic foundation of a free-market society, which is supply and demand. It says that if you want more services—government or otherwise—you must increase the available revenues to pay for them. In a global market, the share of available revenues is limited by the competitiveness of a country's products and services. If country *A* holds an indisputable, competitive advantage over the rest of the world for most goods, the country would consistently run a positive trade balance. In this case, conventional theory would apply, and the government could keep the budget in balance with small changes in taxes and spending without jeopardizing the economy.

On the other hand, if the major trading partners were increasing productivity faster than country *A*, there would come a time when the tide of trade *would* change. The foreign productivity advantage would eventually lead to a price and quality advantage for foreign competition. Country *A* would then begin to realize a negative trade balance, a loss of corporate profits and jobs, low tax revenues and savings, and then a federal budget deficit; hence, Bernstein's theory.

In the case of the country of America, we in fact held the productivity advantage, though it was slowly ebbing away, until the early 1970s. Then, a dramatic increase in oil prices caused an increase in our cost of goods, lower consumer buying power, and a relative decrease in global price competitiveness. What ensued was an economic recession. In the years since 1973, the productivity growth of several foreign economies has outpaced the United States by a considerable margin; see Table 1.1.

In 1973, the American public was accustomed to a relatively high standard of living that depended on an established level of government and domestic services. Therefore, government and private spending for services and defense continued to grow at a constant rate. Politicians and many economists believed that our economy would grow out of this recession, and tax revenues would come back in time. In fact, the Reagan Administration cut taxes in the belief that this would create more spending, increase jobs, give tax revenues a boost, and thus eliminate the deficit. The expected result did not hap-

TABLE 1.1 Manufacturing Labor Productivity (Output/Hour)

Country	1970–1980 Percent Change	1980–1985 Percent Change
Japan	6.6	5.9
France	4.9	4.3
Italy	4.9	3.2
West Germany	4.3	3.5
United Kingdom	2.5	5.7
United States	2.3	3.7

Source: U.S. Department of Labor Statistics.

pen. Tax revenues stayed down and the deficit got bigger. The politicians still cannot make up their minds about whether to cut government spending or raise taxes.

The federal budget deficit has, in effect, financed the continuation of our standard of living, at least as it relates to defense, government jobs, and social services. Left alone, the deficit and interest rates will take their toll on the standard of living. Increased taxes to balance the budget would also reduce the buying power of Americans, as would major cuts in government spending because of the loss of jobs and the reduction of services. What the federal deficit means is that the standard of living of Americans must go down no matter what monetary or fiscal policy the government chooses to follow. The standard of living must adjust to the position we deserve among our trading partners based on our relative productivity and competitiveness.

If Bernstein is correct about the root cause of our economic problem being productivity, it would follow that the only real cure for our economic dilemma is to start a national program to improve productivity. Indeed, President Reagan's Commission on U.S. Competitiveness, chaired by John A. Young, President and CEO of Hewlett-Packard Company, identified productivity growth as the first and most obvious of five elements contributing to the decline of U.S. competitive performance and the economic problem. The following is an excerpt from the report of the President's Commission (Young, 1986):

Indicators of Declining U.S. Competitive Performance

No single indicator gives an adequate representation of our nation's competitive performance. The commission identified five trends, and they all point to a declining ability to compete. First, growth in American productivity has been surpassed by that of all our major trading partners. The Japanese productivity growth rate is five times greater than our own. In absolute terms, Japan is more productive than American industry in autos, steel, and electrical and precision machinery. It is no coincidence that these are the industries which the United States has seen the greatest effects of foreign competition.

Second, real hourly wages have remained virtually stagnant since 1973, and they have actually declined in the last five years. Recall that competitiveness was defined . . . as our ability to succeed in world markets while maintaining our standard of living. Our failure to earn increasing real income means we are not meeting that test.

Third, our manufacturing sector is not generating the kind of real returns on assets that encourage investments. Twenty years ago the average real pre-tax return on manufacturing assets was almost 12%. In 1983, it averaged about 4%. Investors can do a lot better by putting their money in financial assets. The members of the commission were firm in their conviction that we cannot rationalize the poor performance of manufacturing by arguing that we are becoming a service economy, anyway. Our manufacturing sector is the foundation on which many services rest.

The fourth trend that concerned the commission is even more dramatic; U.S. trade deficits are at all time highs—more than $125 billion in 1984. For the entire century—until 1971—we ran a positive balance of trade. Since then there has been a steady—and alarming—trend to the negative. Much of our current deficit can be blamed on the strength of the dollar, but that does not explain it all. Our trade deficit started in the 1970's when most people thought the dollar was 20 to 30% undervalued.

The fifth and final warning signal I would cite hits close to home. Since 1965, seven out of ten U.S. high technology industries have lost world market share. In 1984 this country had a trade deficit in electronics. Our bilateral deficit with Japan in electronics was $15 billion dollars. That is more than our bilateral deficit in autos. Silicon Valley is not so far removed from Detroit.

In assessing our ability to compete, we should not take comfort from the fact that our economy is out-performing the European economies. That is like congratulating ourselves for finishing a race second to last. Instead, we should look to Japan and its neighbors— the new industrializing nations of the Pacific Rim. The United States now does more trade with the countries of this area than with all of Europe combined. And our new Pacific Rim competitors have set a challenging standard by which to judge our own performance.

What can we do to reverse the competitive erosion of the past two decades? It would be nice if we could say, "just do this, and everything will improve." But our ability to compete depends on many factors—all of which are interrelated.

Young's report goes on to define broad areas of national policy needed to influence America's competitiveness. These initiatives call for cooperative efforts between government, business, and education to influence technology, capital resources, human resources, and international trade. This report specifically recommended a cabinet-level position on trade, which was subsequently rejected as being politically unfeasible. Other aspects of the report relate to national awareness and motivation to develop the resolve to compete, in effect, to develop a national program aimed at increasing the nation's productivity.

A national program to improve productivity would have far reaching implications for the federal government. Government would need to become more involved in industry, providing business with the proper level of support and incentives for improving productivity. It would probably mean establishing a cabinet-level position for industry and trade as was recommended by the President's commission. Government, business, and education would have to work together to build a new quality and productivity culture. There are many aspects to national policy that must be attended to, however, they are beyond the scope of this book. There have been many studies done on this topic in recent years, including the President's Commission on Competitiveness. Interested readers are directed to *The Positive Sum Strategy* (Young, 1986).

NATIONAL CULTURE AND PRODUCTIVITY

National productivity is also influenced by national culture. The parts of our culture that affect productivity are our attitudes about energy conservation, waste, quality, individualism, and management. These are the kind of attitudes that are built up on a set of beliefs developed over many years. They are not easy to change, and yet our future economic survival demands that we change them.

Most American people tend to see things in terms of what the country can do for us as opposed to what we can do for our country or ourselves. Indeed, when John F. Kennedy said, "Ask not what your country can do for you, but what you can do for your country," he was addressing the cultural inclination of Americans to depend too strongly on the government to satisfy personal needs.

Our culture also leads Americans to have a high tolerance for waste. Blessed with an abundance of resources, conservation has never been a real need except in recent years, when some resources have come into short supply: like oil and fuel, and fuel-dependent products like aluminum. Consequently, we have never learned to value quality as a key to minimizing waste in the things we do.

When our culture is carried into the workplace, it becomes our business culture unless the business takes some overt action to change it. Selfish motivation, waste, and apathy toward quality combine to detract from the potential productivity of American business. Some businesses can and do establish strong cultures oriented toward productivity. Establishing a strong quality and productivity culture, however, would be easier if it were already ingrained in our society as it is in Japan. Strong national leadership to influence a national resolve for quality and productivity would pay large dividends in national productivity.

The Japanese people have lived with shortages of resources for hundreds of years. As a society, they have evolved a culture that is intolerant of waste in every aspect of their lives. They use every part of the fish they catch; they use every splinter of wood in construction; they use very little space in their homes; they use public transportation to commute; they use fuel-efficient factories, homes, and automobiles. Waste is an intolerable part of their society, and they have learned to control it through quality processes in every part of their lives including the workplace.

Japanese society also has great reverence for harmony as taught in Shinto and Zen Buddhism. Living in a crowded country, the Japanese have acquired a high regard for self-sacrifice, hierarchy, and teamwork as a means to maintain order among the masses. Families and communities work in harmony and teamwork to help one another, as demonstrated when the oldest son or the most gifted child is aided to succeed in business or education.

In the workplace, the Japanese work in quality circles, share goals for process improvement, and enjoy reward systems like profit sharing to reward the team effort. They are in business to do business, and they are not easily tempted to forsake their capital investments when they see alternative short-term investment opportunities. They share management roles through teamwork, cross training, and consensus. They use statistical quality control and structured analysis for improving quality and eliminating waste. They have an appreciation for the complexity and interrelationships of business processes. This structure of thinking is reflected in their planning systems and methodical approaches to customer satisfaction. Indeed, Japanese manufacturing quality and productivity are very deeply rooted in their culture.

In America, the culture is mixed and not quite so frugal. Born of a fight for independence, individual rights, and equality, Americans became a very self-sufficient society. We had a wealth of resources to help build a strong and prosperous economy. The immigrants to America were pioneers who had endured many hardships to carve out a life for their families. A society developed where all people had the opportunity to grow from rags to riches; all it took was hard work, perseverance, and the right opportunity.

During the first one hundred years, Americans moved west and started many new cities, establishing western civilization, capitalism, democracy, and trade across the prairies and the mountains of the continent. Americans developed pride in their achievements. They worked long and hard to establish farms, retail shops, and service businesses. In spite of the hard and often dirty work, most Americans still maintained dignity in the way they earned a living.

Things began to change as the industrial revolution developed. As people moved from the farms to the factories, many of the new jobs took on the appearance of slavery, where people worked in sweatshops. People often had to work long hard hours in filthy and often unhealthy environments to make a minimal living. Pride in job and workmanship began to fall by the wayside. Labor unions were established to protect the rights of workers from uncaring employers. Employers became more set against workers and their unions, not giving any more than necessary. It became a "we and they" society: workers and management.

With the growth of industry also came new technologies, an expanding economy, more jobs, and new ideas. Some industrial leaders began to reject the prevailing practice of human sacrifice in the factories. People like John H. Patterson, founder of the National Cash Register Company (NCR), advocated the rights of American workers to better pay and working conditions. Stanley C. Allyn, one-time employee of John Patterson and later CEO of NCR, wrote in later years: "Patterson had insisted that men and women who worked in factories were not sub-human slaves, but citizens who deserved the best for themselves and their families—even at the sacrifice of some of the profit. Such an attitude was outrageous in 1913." Allyn credits Patterson with many innovations of modern business, including adult education for workers, the incentive quota system, medical care, and noontime recreation, none of which were known to industry at that time. To Patterson, these were simply good business practices and the logical cure to unmotivated and unproductive employees (Allyn, 1967).

In the years since 1913, many businesses have worked their way up the ladder of success. Some, unfortunately, did not survive and fell by the wayside. The survivors are the businesses that G. Harry Stine refers to in his book (Stine, 1986). They are the companies that consistently stayed in the *Fortune* 100 over the years. They survived because they led the race for productivity improvements in their environments. They used *innovation, technology, capital formation, and human resources* to constantly increase the *variety* and sales of their products. Some of the survivors are General Motors, Ford Motor Company, Standard Oil, and Goodyear Tire and Rubber. Along the way, the older survivors were joined by newcomers from the high-tech industries like IBM, Hewlett-Packard, and Digital Equipment Company, among many others.

In their own way, each survivor contributed in some unique way to the development of our national business culture. Some companies, like NCR, IBM, and Hewlett-Packard, developed corporate values and cultures that are characterized as the best human resource models in the industry. Some excel in quality, others in product innovation, and some in financial management and control. Most of these companies excel in several areas and are considered working models for business schools and academics to create training programs for future business leaders.

The survivors are certainly the best run businesses in America. They are also among the largest businesses in America. Unfortunately, their success alone cannot support the U.S. economy. There are thousands of other companies who supply them with parts, and thousands of others who are customers. Collectively, all of these businesses are part of the "macrobusiness system."

Sometimes managers forget that their business is part of a larger system of businesses. This macrobusiness system consists of the many suppliers and customers and their interrelationships. An individual business indeed may be as well managed as the survivors are, but it is the performance of the macrosystem that determines our national competitiveness. Each business is a supplier and a customer in this system. The quality, cost, and delivery performance of the supplier has a large impact on the productivity of the customer.

An example of these relationships comes from the Hewlett-Packard Company. In 1980, CEO John A. Young established a "stretch objective" for the reliability of HP's products. This objective called for a ten times improvement in the customer failure rate over the next ten years. Now, during the ninth year, it is apparent that the desired improvement is mainly on track. Where did the improvements come from? The answer is not simple, and maybe even debatable to some extent, but several factors contributed to the improvement. First, the design labs took reliability much more seriously in their designs, testing, and component selection. Second, manufacturing became more aggressive in the pursuit of product and process problems in production. Third, and just as important, the reliability of the components improved considerably through technology and process improvements of HP suppliers.

Every large manufacturing business has hundreds of suppliers and customers. Every one of them is mutually dependent on each other for its success. If U.S. manufacturing is going to make a significant improvement in national productivity, then efforts must be made to manage the macrosystem as well as the individual factory. This could be the function of industry associations if properly funded and chartered by business. It could also be the purpose of a cabinet-level position on industry and trade. The national cost of not maintaining industrywide leadership in cost, quality, and delivery is far too significant to be ignored.

At this time, the competitiveness of U.S. manufacturing is certainly below what it was fifteen years ago. There is also sufficient evidence to support the belief that the U.S. economy is suffering because of this loss of competitiveness, and that low productivity growth is a leading contributor. This would indicate that unless the United States becomes a large exporter of services, the future of the U.S. standard of living is still in the hands of the people who manufacture things.

REFERENCES

Allyn, Stanley C. 1967. *My Half Century With NCR*. New York: McGraw-Hill.

Coy, Peter L. 1988. "Productivity Still the Economic Key." Associated Press, January 8.

Ford, Henry. 1922. *My Life and Work*. Garden City, NY: Garden City Publishing.

Hudson, William J. 1987. *Business Without Economists*. New York: AMACOM.

Roach, Stephen S. 1986. "Macrorealities of the Information Economy." In *The Positive Sum Strategy*, edited by Ralph Landau and Nathan Rosenberg. Washington, DC: National Academy Press.

Stine, G. Harry. 1986. *The Corporate Survivors*. New York: AMACOM.

Young, John A. 1986. "Global Competition The New Reality: Results of the President's Commission On Industrial Competitiveness." In *The Positive Sum Strategy*. Washington, DC: National Academy Press.

2

Manufacturing Productivity

During the early years of the industrial era, Henry Ford set the standard for manufacturing and productivity with the Model T, using mass production. There were many benefits associated with mass production for both business and workers. Businesses became more productive and therefore more price competitive. At the same time, higher productivity led to higher pay and shorter workdays for employees (Litchfield, 1954). As mass production spread throughout the industry, manufacturing strategies became centered around a "bigger is better" philosophy. Many manufacturers narrowed their product focus and increased the volumes as much as possible, using large lot sizes to take advantage of economies of scale.

Mass-production strategies worked very well during the first half of this century. As the second half of the century developed, however, Japanese competitors began to introduce new manufacturing concepts, many of which were developed in the United States. At first, Japanese manufacturers competed with the same mass-production strategies used in the United States, however, the Japanese were more competitive because they had lower wage rates. Later, Japanese manufacturers began to focus on quality and capital-investment strategies. This focus led to stronger productivity growth and enabled Japanese competitors to dominate many U.S. markets.

In the late 1970s and early 1980s, Americans recognized the value of the Japanese tactics and began to put more emphasis on quality and, more recently, capital investment. While Americans were plowing their way through books on Japanese total quality control, the Japanese had already begun to further enhance their strategy. More competition and more demanding customers were establishing a need for a wider variety of products. It was possible for large mass-production companies to be picked apart by many niche competitors. New manufacturing technology was making it possible for small companies to be price competitive with large mass-production manufacturers.

The Japanese responded both defensively and opportunistically by offering more product variety to their customers. Variety in scale-based manufacturing, however, is more costly because of the requirements for more inventory, more setups, and smaller lot sizes. To overcome this added cost, the Japanese began to increase factory flexibility by reducing throughput time and setup times (Stalk, 1988).

TIME-BASED MANUFACTURING

George Stalk, Jr., refers to the Japanese move to greater variety and short throughput time as "time-based manufacturing" (Stalk, 1988), as opposed to "scale-based manufacturing." Scale-based manufacturing depends on the economies of scale that result from a low variety of high-volume products. Time-based strategies are based on high-variety, fast response, and flexible manufacturing processes. Stalk suggests that time-based manufacturing captures the benefits of scale-based manufacturing with high product variety by eliminating or reducing the time-consuming tasks that result from frequent changeover costs of lower-volume, high-variety manufacturing.

Stalk uses two illustrations to make this point. Figure 2.1(A) illustrates that cost comes down as volume increases and goes up as variety increases. Higher variety at a given total volume reduces the volume on individual products. In effect, the learning curve, setup, and inventory requirements become a larger total part of the product cost. If a business can reduce the time and cost associated with higher variety, in effect, it reduces the slope of the variety/cost line; see Figure 2.1(B). When this is done, the optimum volume/variety point is moved to the right, providing the business the benefit of higher variety at a lower cost.

If the costs of variety could be taken to zero, the slope of the line would be zero and the business could operate at infinite variety without a loss of productivity. This is a theoretical impossibility. The real

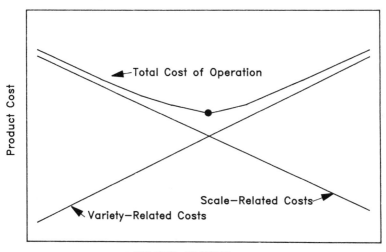

FIGURE 2.1(A) Cost vs. volume and variety in traditional manufacturing

SOURCE: Reprinted by permission of *Harvard Business Review*. An exhibit from "Time—the Next Source of Competitive Advantage" by George Stalk, Jr., July/August, 1988. Copyright © 1988 by the President and Fellows of Harvard College; all rights reserved.

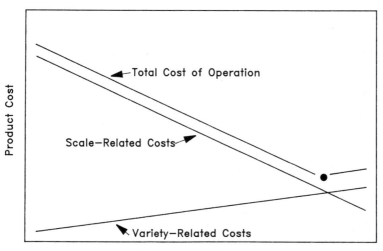

FIGURE 2.1(B) Cost vs. volume and variety in flexible manufacturing

SOURCE: Reprinted by permission of *Harvard Business Review*. An exhibit from "Time—the Next Source of Competitive Advantage" by George Stalk, Jr., July/August, 1988. Copyright © 1988 by the President and Fellows of Harvard College; all rights reserved.

cost relationship to both volume and variety is an exponential relationship, as shown in Figure 2.2. The volume curve is a product of the investment profile, learning curves, setup time, and volume. The variety curve is due to the multiplicity of the investment profile, setup times, learning curves, and lower volumes of a high variety of products. The variety curve can be reduced by minimizing the duplicate investment required through the use of generic and flexible equipment, lowering setup costs, decreasing the learning curve, and using nonpenalizing inventory systems.

The goal is to push the variety curve down and move the knee of the curve to the right as far as possible and operate on a relatively flat part of two exponential curves. When this is done, a range of cost and variety is created, whereby a business can operate with only a small change in productivity. A factory must be designed so that it comes as far down the volume–cost curve as possible and have the range of flexibility needed to compete in its industry. Further, it must operate within that range to retain a competitive position.

A key part of time-based manufacturing strategy is the just-in-time (JIT) manufacturing system. When JIT principles are applied from

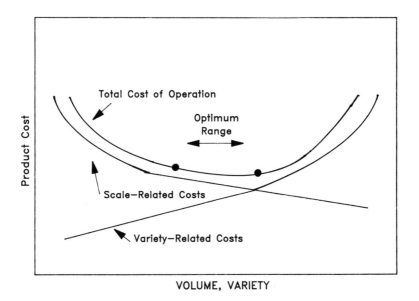

FIGURE 2.2 Optimum cost vs. volume and variety

supplier delivery to customer order delivery, the overall cycle time (throughput time) is reduced. Studies demonstrate the statistical correlation between reduced cycle time and productivity improvements (Schmenner, 1988). The productivity improvements come from eliminating the quality problems and inventory that drive labor and overhead costs in manufacturing processes. The results of implementing JIT and reduced cycle-time manufacturing vary between businesses, but the results are consistently positive. On average, businesses have achieved the following benefits (Coleman, 1988):

58% reduction in in-plant inventory

60% reduction in material-handling costs

20% reduction in indirect labor costs

30% improvement in delivery times

The benefits shown here are just a sample of what can be achieved. In January 1989, *Fortune* Magazine published a report on the benefits and progress of managing speed (Dumaine, 1989). This report looked at the benefits of speed applied to both production and innovation. My own company and division (Hewlett-Packard, Lake Stevens Instrument Division) were cited in this article for 80% improvement in production cycle time and an 80% reduction in work-in-process inventory. Not mentioned was the 30% reduction in labor costs and an overall improvement in production cost of 25%. These improvements came about from a manufacturing focus on three strategic areas of focus: quality, cost, and cycle time.

THE PRODUCTIVITY PARADOX

Production cost, quality, and cycle time are directly influenced by the design attributes of the product and the manufacturing process. Product attributes change, depending on the application, industry, and technology. For a given product, the production cost may vary widely between manufacturing sites, depending on local management's choice of manufacturing strategy.

Some manufacturing managers do not know where to start looking for productivity gains. They get caught up in the belief that reducing direct labor is the primary path to productivity improvements. This has led to what Wickham Skinner calls "the productivity paradox'' (Skinner, 1986).* In the attempt to increase output or reduce labor, businesses put increasing emphasis on programs designed to improve the efficiency of direct-labor employees. The paradox arises when managers invest heavily in cost-reduction programs. They do efficiency analyses, simplify jobs, install automation, and strive to eliminate waste. Later, they find that their productivity and competitive position changed very little. In some cases, the productivity may even decline because the small gains in direct labor costs are offset by large increases in overhead.

Labor, it turns out, is a small part of the total cost of sales in most businesses. As Skinner reported, it is less than 10% for many businesses, and specifically in electronic manufacturing, it is less than 5% of sales. The large components of manufacturing cost are material and overhead, which may combine to be 40% of sales in many industries. Reductions in these areas require a much more strategic involvement on the part of manufacturing in the design of products, processes, and investments.

Breaking out of this paradox, Skinner reported, requires businesses to make "changes in their culture, habits, instincts, and ways of thinking and reasoning." Based on his study of twenty-five companies, Skinner observed five success factors associated with breaking away from the paradox created by traditional thinking:

1. Recognizing that its approach to productivity was not working well enough to make the company cost competitive.
2. Accepting the fact that its manufacturing was in trouble and needed to be run differently.
3. Developing and implementing a manufacturing strategy.
4. Adopting new process technology.
5. Making major changes in the selection, development, assignments, and reward systems for manufacturing managers.

MANUFACTURING STRATEGY

A quick look at the current economic environment should allow us to move quickly through steps 1 and 2. We will move directly into step 3 and begin to develop a manufacturing strategy.

In developing a manufacturing strategy, it is important to understand manufacturing in its simplest form—a pipeline. In the manufacturing pipeline, material is funneled into the pipe from suppliers. Value is added all along the pipeline at each process step. At the other end of the pipe, a finished product emerges and is shipped to the customer. Ideally, the time from suppliers to customers would take no longer than the sum of the times for the sequential value-added process steps. If this could be achieved, the cost to manufacture the product would be no higher than the sum of the costs of each value-added process step. These costs would include the labor costs, equipment expense, space cost, and the cost of storing process information. This is the theoretical minimum cost.

Compare the theoretical minimum cost to the actual cost. In most manufacturing operations, the actual cost of a product is two to three times higher than the theoretical cost because:

1. poor-quality parts are being introduced into the pipeline;
2. additional defects are created in the processes;
3. queues of inventory require storage and tracking;
4. long supplier lead times require hundreds of open purchase orders and tracking systems;
5. poor supplier delivery performance requires a horde of buyers to continuously adjust to changes and expedite parts;
6. poor designs require a phalanx of manufacturing engineers to fix problems;
7. all the quality problems require defect-tracking systems;
8. the complexity of manufacture requires a master planning system (MPS), materials requirement planning (MRP), work orders, and many other paper work and system-tracking systems;
9. systems and the like require large investments in programmers and equipment.

These added manufacturing pipeline "overhead" costs are just a sample of the costs that burden present-day factories.

The problems basically have four sources:

1. the quality and flow of the material from suppliers;
2. the quality of the design of the product;
3. the quality of the process technology (hardware);
4. the quality of the information flows (software).

These quality problems act as obstacles to the continuous flow of material along the pipeline. By choice and necessity, pools of inventory (material queues) and backlog (unfilled order queues) build up along the way. Left unattended, these quality problems eventually build up and shut off the flow through the pipeline; therefore, extra engineers, programmers, process analysts, and managers are required to continuously break up these logjams. The human problem solvers require sophisticated systems to make their problem-solving tasks easy, though the systems contribute nothing to the value of the product. The throughput time, therefore, becomes the sum of the total value-added process time, inventory queue time, and backlog queue time. The throughput time then becomes an *indicator* of the number of quality problems with which a factory has to contend and therefore an *indicator* of the amount of "overhead" required to keep the processes flowing.

Consider, as an example, how a manufacturing factory operates and how it can be improved. There are six major manufacturing functions to be considered:

master production planning

materials requirement planning

procurement and stores

production control

production and shipping

manufacturing engineering

Master Planning System

Master production planning is normally done with a master planning system (MPS). The MPS combines information from the order-processing system and the sales forecast to create a production schedule. The production schedule is adjusted to provide linear weekly or daily production volumes as required to meet the customer delivery requirements. The MPS also lets the order coordinators know when to quote delivery on a product.

Some companies like to operate with a backlog (unshipped orders), some with finished goods inventory (FGI), and some operate with both. The backlog and FGI are necessary because the production cycle time is long and the business has limited capacity flexibility. The existence of backlog means that customers who order a product today cannot receive it until all prior orders are shipped. This may take several weeks, depending on the level of backlog. The presence of FGI requires unnecessary space and storage equipment.

Materials Requirement Planning

The output from the MPS is used to drive the materials requirement planning (MRP) system, which is used to tell production control and procurement how many work orders and purchase orders to prepare and send out.

When the system is put into motion, material arrives with enough safety stock to fill up the warehouse and meet the plant's work-in-process (WIP) inventory requirements. Labor is soon added and the desired level of FGI is established. Customer orders are now being processed and shipped, and the system is functioning as expected. Teams of buyers, schedulers, engineers, systems programmers, administrative support people, supervisors, and assemblers are doing their job to keep the system operating and improving.

The financial profile for this example business is defined as follows. This business is referred to as XYZ Manufacturing in future paragraphs.

Financial Profile of XYZ Manufacturing

Production cost percent of sales:	45–50
Labor cost percent of sales:	4–8
Material cost percent of sales:	20–25
Overhead cost percent of sales:	15–20
Customer delivery time:	6–10 weeks
Production cycle time:	6–10 weeks
Supplier lead time:	10–16 weeks
FGI:	2–3 weeks supply
WIP inventory:	6–10 weeks supply
Raw parts inventory:	16–20 weeks supply

XYZ Manufacturing is typical of many traditional manufacturing facilities. Specifically, the profile fits many electronic assembly businesses, but is not too far from describing factories in other industries. This makes XYZ Manufacturing a good candidate for applying a manufacturing strategy suitable for the "factory of the future," which addresses quality, cost, and time as the strategic elements.

The strategy for XYZ Manufacturing is pay as you go. Continuous improvement is seen in manufacturing cost, investment levels, and delivery times. For example, continuous quality improvements contribute to lower labor and overhead costs; lower inventory and inventory handling costs more than offset the investment and support costs of automation. When the strategy is implemented as described here, quality, cost, and delivery performance get better from the first day and never need to be justified based on future returns.

There are three phases to this strategy. Each phase brings the business to a higher plane of competitiveness. Phase 1 is focused on the internal workings of the production environment, and phase 2 is expanded to include the relationships between customers and suppliers. Finally, phase 3 addresses the productivity of the entire organization, which includes marketing, design, and administrative functions.

Phase 1

There are five steps in phase 1. Each step builds upon the other, though there is overlap. The steps are purposefully arranged to achieve the greatest gains in productivity, the fastest implementation, and the greatest return on investment to the business.

Step 1.   The first step of this strategy is to address the quality of the processes through a total quality control (TQC) system. The TQC system addresses the quality problems described before as inhibitors to throughput and productivity. The specifics of TQC are discussed in detail in later chapters. But, suffice to say here, TQC needs to become thoroughly embedded and "institutionalized" in the business as a foundation for all future strategic improvements.

The purpose of the TQC system is to gain control of all business processes and establish a framework of continuous improvement aimed at satisfying customer needs. Motorola's "Six Sigma Quality Program" is an example of a world-class approach to TQC (see Appendix A). By achieving six sigma improvement in all of their processes, Motorola is virtually assured a position equal to the best manufacturers in the world. From a customer's point of view, few if any will be dissatisfied with the mean time between failures of Motorola's products. From Motorola's point of view, it will realize reduced costs, shorter cycle times, and better availability, which will bring it unparalleled growth and profitability in its industry.

After the organization has become adept at managing quality and controlling the improvement of its processes, it can proceed to the next step. The next step after TQC is production JIT.

Step 2. XYZ Manufacturing will implement a JIT kanban (card) pull system in their production operation. The JIT system will reduce the cycle time and WIP requirements in the production process. The production pull process begins when a product is pulled from FGI to ship to a customer. This pull process causes production to initiate another product build cycle by releasing kanban cards to the printed-circuit assembly area and thus start a build sequence (see Appendix B). The build cycle ends when the product is passed to finished goods inventory.

JIT works to improve cycle time because the process variables are reduced to the lowest possible levels. The unnecessary queues are removed; therefore, each piece of inventory spends less time at each spot. This relieves the manufacturing process of having to contend with excess inventory. The assembly people are able to do this without a loss of productivity because they have access to improved documentation and part-handling equipment. Higher levels of quality are a requirement for parts going to a process and a benefit for those leaving a process, thereby reducing rework loops. JIT also improves the coordination of part and subassembly flows at the assembly level. Pull systems as opposed to push systems provide this, while simplifying shop-floor tracking, work prioritization, and decision requirements. Fewer inventory transactions are required throughout the process with kanban JIT processes.

JIT implementation requires process analysts to understand the product structure, assembly processes, and flow of material through production. The simplicity of kanban procedures removes the need for complex systems and the need to centrally understand the part quantities required at each assembly station. The need to prioritize material pulling and assembly is also eliminated; the pull system automatically does this. In effect, a kanban JIT system makes manufacturing process decisions simple and obvious and thereby reduces time.

Quality and cycle time work hand in hand to improve production cost. Quality problems contribute to the reasons for high inventory and long throughput times, which in turn contribute to high overhead costs. A popular analogy relates to lowering the water in a lake that has many submerged rocks. When the water is lowered, some of these rocks appear at the surface. More rocks continue to appear as the water is lowered further. If a boat were to attempt to race across the lake as the water is lowered, it might hit one of these rocks, so it must slow down to maneuver around them. The lower the water, the slower the boat must go. The water is analogous to inventory; it hides many quality problems. When inventory is lowered, production may stop unless these problems are eliminated. After the quality problems are eliminated, the inventory can be eliminated and the throughput time is shortened.

Short cycle time requires that work be done right the first time. When this is accomplished, scrap, internal and external defects, rework, labor, and the associated overhead expenses decrease. When inventory goes down, space is saved and material-handling labor is lower. The benefits of the TQC and JIT implementation are as follows:

Benefits of Production JIT and TQC

1. Internal and external defects are reduced by 70–80%.
2. Production cycle time is reduced from 6–8 weeks to one week.
3. WIP is reduced to a 1-week supply.
4. Production labor efficiency is improved by 10–20%.
5. Material-handling and production-control expenses are reduced by 50–60%.
6. Production space is reduced by 20–25%.

Step 3. Part of the savings from the initial implementation of TQC and JIT will be reinvested in higher levels of automation for the manufacturing processes. State-of-the-art automatic insertion equipment for the printed-circuit assembly shop will be purchased. In some cases, this equipment will be fully automated, and in others, it will be computer-aided hand loading equipment. Computer-aided assembly documentation will be set up to provide assemblers with on-line assembly information. As the new technology is brought into operation, further improvements in cost and quality will be seen. Defect rates and assembly costs will be approximately one-fifth to one-tenth what they were for traditional hand operations.

Step 4. The next step is to integrate the automation and people in the processes through efficient information flows. Computer information management (CIM) is a key element in world-class manufacturing. Learning curves and setup times, for example, can be reduced by simplifying and automating the information flows required to change from one part to another. Consider the comparisons between the following hypothetical technologies in a shop that makes a variety of parts. This comparison is of two different process technologies (see Table 2.1) making the same part with the following characteristics:

• one work order per month of 200 parts
• requires ten process steps

In this simple example, the production manager could improve the result of process A by doubling the work order quantity and operating farther down the learning curve. This would make it possible for process A to be more cost effective than process B. In doing so, however, more capacity would be used up by one part type and the flexibility to make a variety of parts would be lower. Also, inventory and/or delivery times would be higher, space requirements would go up, and scrap levels would increase unnecessarily. The manager faced

TABLE 2.1 Two Different Process Technologies Making the Same Part

	Technology A (People)	Technology B (Robot)
Instruction set	book/memory	computer memory
Format/media	learning curve 75%	continuous
Process/speed	0.4 steps/min @ 1	no learning curve
	5 steps/min @ 200	3 steps/min
	avg. 2.0 steps/min	
Setup time	0	0
Decision-making time/step	0.25 min	0.1 min
Process accuracy	99%/step	100%/step
Yield	90%	100%
Total time/job	2000 steps/2 + 500 min	2000 steps/3 + 200 min
	= 1500 min	= 867 min
Cost/min	0.40	0.50
Yield	180 parts	200 parts
Cost/part	$600/180 parts	$433.5/200 parts
	= $3.33/part	= $2.16/part

with this process decision might also consider a third alternative: use process A combined with on-line documentation.

On-line computer documentation provides up-to-date illustrations, with exploded views and step-by-step instructions, all on a computer screen directly in front of the operator. Using on-line documentation allows employees to rotate in and out of jobs and assemble and test different products frequently without a loss of quality and time that results from learning curves. It also requires less supervisory support. Experience demonstrates that the benefits of on-line documentation are

1. Shorter learning curves (one-third to one-tenth the normal learning times);
2. documentation is easier to keep up to date with production changes;
3. the employee can make faster decisions about the quality of each step (10:1 reduction in the number of defects).

On-line documentation allows employees to handle assembly and test information more accurately and in less time.

On-line documentation may be the only avenue available to processes where technologies do not support higher levels of automation. If this were the case in the last example and on-line documentation were made available, the part cost and quality of process A could be equal to or better than process B. This can be verified by reducing the learning curve by 2:1 and recalculating the part cost based on the new average times.

On-line documentation, however, is just one imaginative approach to reducing learning curves through information management. A little imagination and creativity can go a long way in identifying untapped opportunities for applying information technology. In many cases, it is not necessary to wait for advancements in technology; it exists today. It does, however, require a change in thinking. The people in the organization must begin to think of every task in terms of its information content and processing time.

The business telephone system is another good example of applying information technology. Most telephone calls require the person making the call to look up the number. In a typical business day, 80% of those calls are repeated monthly, but not often enough to remember. Comparing the cost benefits of a programmable telephone that stores 200 numbers by name, combined with automatic dialing, to a standard phone illustrates this point; see Table 2.2.

TABLE 2.2 Cost Benefits of a Programmable Telephone

Standard Telephone		Programmable Telephone
Telephone cost	$60.00	$120.00
Look-up time	30 s	0
Number calls/year	2000	2000
Look-up time/year	1000 min	0
Cost/min	$0.50	$0.50
Cost/month	$42	0
Payback time	—	$60/$42 = 1.43 months

This simple example may not, however, be justification to buy every employee a new telephone, since it may not be the biggest payback item a business can find. Each business has an operating budget with which to contend and must set priorities accordingly. It may also be that the business can achieve the same result by integrating the phone directory and automatic dialing into its voice-mail system. Since voice mail requires a network with some amount of central processing, the company's phone directory can be set into a data base and provide automatic dialing keyed from a name. Adding the ability to allow employees to program outside numbers would greatly enhance this system.

The previous examples illustrate that there is usually more than one technological choice to reduce the time in a process. The final choice of technology must be based on the level of investment and integration possible within a given business.

The concept that all manufacturing activities are "information-handling processes" provides a useful model for developing manufacturing strategies. Focusing on the information-processing content of fabrication, assembly, administration, and engineering provides a common framework for thinking about technology and strategies. To use the pipeline analogy again, think of all value-added work along the pipe as being performed by humans, machines, robots, or some combination of the three. As processors, they are the hardware and require an information-distribution strategy (software) to perform their jobs at high-quality levels in optimum time.

Information strategies must consider the information commonalities in the different technologies. Humans and robots, for example, both require a set of information describing what steps are to be done and in what order; both are expected to make decisions about the completeness of a step before proceeding; both may be asked to process visual information on the work in progress, make decisions about quality, and take compensating actions; both are usually expected to operate at some predetermined measure of speed and quality. As processors, both humans and robots are continuously assimilating information and translating it into some form of action.

Humans and robots differ in their relative level of intellect, among other things, but they also differ in the language and media used to communicate information. The requirements for language and media are determined by the process technologies employed to do the work and/or take action. (Work is defined as interpreting a set of information, making decisions about it, and taking some action to create or add to the value of a product or service.) Humans require information that can be interpreted by the senses (vision, touch, sound, etc.) and processed by the brain, which makes a decision. An example is the on-line computer (visual) documentation discussed before. Robots require machine-readable information that can be interpreted by sensors (video cameras, microphones, physical measurements, etc.), processed by a computer, which makes a decision, and then implemented into action. In the case of a robot, the media is electromechanical, and the language is formed from binary-coded programs.

There are many levels and types of technology. The biggest challenge manufacturing managers face is knowing when and what new technologies are available and whether they offer savings over existing methods. The common element of comparison in information-processing technologies is time. As managers learn to use information technologies to better manage time, they will find that fewer managers are necessary, organizations will become flatter, with more centralized control, but with more decentralized decision making. But, these benefits will not be limited to management; all information-processing jobs will take less time and, therefore, be fewer in number.

Step 5.   People and organizations present another opportunity for saving time at XYZ Manufacturing. The more complex the organization, the more time it takes to develop shared plans and congruent goals, arrive at consensus and make decisions, communicate and implement plans. Organizational complexity is driven by (1) the number of levels of management in the hierarchy; (2) the breadth of responsibility and number of managers on a lateral basis; and (3) the relatedness of process responsibility under a given manager. Strategies to reduce organizational complexity and the related information-processing time must address each of these elements directly.

The concept of self-managed teams is a step in the direction of organizational simplification. Self-managed teams provide for decisions to be made at the lowest level by the people closest to the process. Teams are focused on their individual process and assume full responsibility for improving it. Less supervision is required and, therefore, managers can take on a greater span of control. Coupled with improved quality information systems, JIT, and other process-simplification efforts, it becomes possible to shrink the size of the management structure and save time in making decisions.

Phase 2

The previous five steps are only the first phase of building a factory of the future. They represent the greatest immediate opportunity for improving quality, cycle time, cost, and availability. However, there are many other functions to be considered in a complete time-based manufacturing strategy. The improvements already seen in the example business are the following: cycle time is at 1 week, production labor and overhead are lower, and quality is great! Now what?

Besides the production component of productivity, there are also the order-processing, procurement, stores, system, engineering, and administrative costs of manufacturing. Throughput time in these processes also contribute to XYZ Manufacturing's overall productivity.

Build-to-Order System. The next step is to look at order processing and the MPS. If XYZ Manufacturing is converted to a "build-to-order" company, there are many potential savings. XYZ Manufacturing operates at a 6–8 week backlog and carries FGI because of the 6–8 week manufacturing cycle time. The cycle time has now been reduced to 1 week. This provides the opportunity to start building a product when the customer order is received and deliver it 1 week later. Now XYZ Manufacturing can build anything, limited by a range of flexibility and capacity within 1 week. Therefore, the 2-week supply of FGI can be eliminated and backlog can be reduced to 1 or 2 weeks. Additionally, the order coordinators no longer need to look at the MPS to quote delivery dates to customers. Instead, they will look at capacity constraints and load the production line accordingly. They will do this by controlling the kanban releases to the production line. Moving to a "build-to-order" system eliminates one use of the MPS. It also eliminates 2 weeks of inventory. Further, the order coordinators spend less time manipulating changes in delivery schedules.

Benefits of the Build-to-Order System

1. FGI space and capital absorption is reduced;
2. order coordinator overhead is lowered by 20–25%;
3. on-time delivery performance to customers is improved by 25–30%;
4. delivery time to customers is reduced from 6–10 weeks to 1–2 weeks.

XYZ production is now building to order and has established a flexible manufacturing environment to respond to short-term changes in orders. The kanban system combined with the build-to-order system has eliminated the need for production control to write work orders and control the MPS. At this point, *the production-control department is no longer necessary.*

Benefit of Kanban and Build-to-Order System

1. Eliminate production-control function, saving overhead costs.

Procurement

The next step is to look at the procurement function. Procurement now has the responsibility to ensure that there are enough materials available to meet the short-term mix and capacity changes. Currently, the business has 16 weeks of inventory, and supplier lead times are less than 16 weeks. This should enable procurement to trigger purchase orders weekly based on the previous week's usage, eliminate the need for an MPS, and simplify the MRP system. The MRP will now only monitor stock on hand, the "bill of materials," supplier lead times, and long-term forecast changes. Removing the MPS will reduce the complexity and cost of running the MRP. The MPS and MRP changes provide additional overhead cost savings by reducing disk space (hardware), programming support, and system administration.

Purchasing must now look at opportunities to reduce supplier lead times and improve on-time delivery performance. Improvements in both of these areas will reduce the amount of parts inventory required. Purchasing should also work to reduce the number of purchase orders written and increase the number of deliveries. This can be done by writing blanket purchase orders to cover a long period of time (quarterly, for example). Minimum and maximum quantities can be stated to ensure that both the buyer and supplier are protected. A payment system based on invoice for delivered quantities should be set up. After purchasing has set up these procedures, they will be in a position to handle JIT deliveries from suppliers.

Benefits of Procurement JIT

1. Parts inventory is reduced from 16 weeks to 3 weeks or less;
2. stock room space is reduced 70–80%;
3. buyer and administrative support costs are reduced by 50%;
4. manufacturing system costs are reduced by 40–50%.

At this time, we have covered the entire manufacturing pipeline JIT strategy. Let us now compare the before-and-after financial profile of the example business; see Table 2.3.

The additional benefit not shown in these numbers is the additional cash flow realized from the reduced inventory and lower capitalization required for land and buildings. This will be several millions of dollars for many businesses.

The benefits of implementing time-based manufacturing shown in the example company are real. The improvements are not based on a given company, but are representative of the benefits that many com-

TABLE 2.3 Example Business

	Before	*After*
Production cost percent of sales	45–50	25–35*
Labor cost percent of sales	4–5	2–3
Material cost percent of sales	20–25	15–20
Overhead cost percent of sales	15–20	7–10
Customer delivery time	6–10 weeks	1–2 weeks
Production cycle time	6–10 weeks	1 week
Supplier lead time	10–16 weeks	2–4 weeks
FGI	2–3 weeks	0
WIP inventory	6–10 weeks	1 week
Raw parts inventory	16–20 weeks	2–4 weeks

*This production cost percentage may never be realized because prices may be lowered to enhance competitiveness.

panies have achieved. It has been through personal experience, and the research of others as reported here, that I have confidence that any business is capable of achieving these results.

Phase 3

Thus far, we have only discussed manufacturing's contribution to improving the competitiveness of the business. Yet, even as much as we have improved, XYZ Manufacturing's productivity is still only better by 50% or so. What we have accomplished so far is simply to do what everyone else is doing. Others are very likely to achieve the same improvements within a couple of years. Then where will we be?

The next phase requires that we improve the overall productivity of the entire organization. This means introducing products faster than competitors that do a better job of meeting customers needs and are inherently lower in cost and of better quality (manufacturable). This is phase 3 and the final phase, or, arguably, the first phase in achieving sustainable competitive leadership.

BREAK-EVEN TIME

In the *Fortune* article referred to previously on managing speed, several companies, including Hewlett-Packard, are reported to have reduced development time by 50% (Dumaine, 1989). John A. Young, CEO of HP, is quoted for his now well-known goal of improving the break-even time (BET) of new products by 50% across all of HP. BET is defined as the time from the start of a new product development cycle to the time of full investment recovery.

The elements of the break-even time equation are the design cycle time and cost, and the ramp-up time, level of sales, and profitability after introduction. Businesses that are successful at lowering BET are more productive and more competitive as measured by revenue growth and profitability. BET cannot be overlooked in the effects of speed on total business productivity.

The productivity of the design lab and marketing is the primary contributor to BET. These functions are also time-based processes and must be viewed accordingly. The same principles that apply to production also apply to design and marketing. They must learn to use information simplicity, technology, and JIT principles in the development and sales of products. It is not my intent to go into these areas as deeply as manufacturing, but some aspects of the philosophy must be covered because of the impact it has on manufacturing.

In George Stalk's article on time-based manufacturing, he refers to the Toyota experience with production and sales (Stalk, 1988). After achieving 5-day manufacturing cycle time, Toyota became frustrated with not being able to make similar progress with the order and distribution time in the sales organization. Eventually, they reorganized to combine Toyota manufacturing with Toyota sales. Out of frustration, the sales directors were subsequently replaced with manufacturing directors. In a short while, the order and distribution time was reduced to 5 days.

Aside from the fact that the Toyota story is good for the egos of manufacturing managers, it also points out that there are still too many managers (manufacturing included) who do not understand or appreciate their contribution to business productivity. They are content to let production people worry about it. A personal experience comes to mind to further illustrate this point. I was attending a meeting that involved heads of several functional departments of a very successful $1 billion per year business. We were analyzing the success factors associated with the most successful business unit. In the course of this discussion, I suggested that their success was derived from productivity improvements: their organizational productivity was higher than their competitors, and their products offered higher productivity to their customers than the competitor's products.

The marketing manager did not agree with this. He argued that productivity was the measure of "doing things right" to improve quality and minimize waste. *Their* success, he argued, was due to having done the "right things" to understand customer needs and develop the right products. Furthermore, he argued, *their* manufacturing productivity was not especially great. In response, I asked him if doing the "right things" did not amount to saving a lot of wasted time on the part of marketing, R&D, and manufacturing; and if doing the "wrong things" did not result in a total waste of everybody's time and equipment. My point was made at that meeting, but there are many others who still do not have that perspective.

In that particular instance, the productivity of the design lab and marketing was higher than that of manufacturing. They had been introducing a series of new products that were well suited to the market. The new products were well defined, used new technology, and had better quality than the competitor's products. As a result, sales skyrocketed and BET was lower than ever before. Profit was high even though manufacturing's performance was mediocre. Why did they do so well even though manufacturing was mediocre? Very simply, manufacturing was forced to respond to unexpected sales and use their resources to the fullest capacity.

This brings up an interesting point. Much of the pressure on manufacturing performance and productivity would be reduced if the design and sales functions were operating at higher productivity levels. This would allow manufacturing to operate at full capacity, providing greater utilization of resources and higher productivity.

Design for Manufacturability

Design for manufacturability (DFM) is an area where design has a large impact on manufacturing cost, quality, and serviceability. This is covered in more detail later, but a short comment is necessary here. Major parts of manufacturing overhead costs are engineering and procurement costs. These costs are driven by the number of different components, suppliers, and design changes required to keep products flowing through the pipeline. DFM must be a key part of every manufacturing business strategy if it expects to achieve the benefits described by the example business.

TOTAL BUSINESS PRODUCTIVITY

Total business productivity is defined by the quality, efficiency, and cycle times of marketing, design, manufacturing, and administration. When more products are introduced that result in more sales for a given investment, then productivity is higher. When manufacturing cost is lower, thus generating more sales through lower price, then productivity is higher. When inventory and cycle times are lower, thus providing a higher return on assets, then productivity is higher.

Management's Role

The management system is responsible for pulling together the labor and capital to work in harmony. Like a symphony, each person and process has a special part to play and must be orchestrated with every other part. The conductor waves the baton to set the beat and call forth emphasis when needed. But within each person playing an instrument lives a skilled musical talent that hears, sees, and feels the music being played. Their actions become a finely tuned network of interaction and harmony.

The idea of business as a symphony is vividly described by Noel Boudette (1989) in *Industry Week* Magazine. Boudette portrays business in the future as a network of people sharing goals and responsibilities. The people are tied together by teamwork and effective communication systems. Tasks are performed by specialists who understand their jobs and apply their talents much like musicians do in performing a symphony. The role of supervising becomes systematic, and requires few managers, much like the conductor of an orchestra.

My experience with teamwork and participative management at Lake Stevens reinforces the conclusions and theories of Boudette. Team members learn to take on extra tasks, based on their preferences and abilities. Some people want to take on extra responsibility for quality reporting and feedback, and others want to have responsibility for material distribution and replenishment. Some people do not want to do anything except assemble their products, and they do it extremely well. When the responsibilities of the teams are well balanced to fit the needs of the individuals and the needs of the business, a high level of quality, efficiency, and harmony is achieved.

As I have progressed with these concepts, I have recognized the need for better communication and decision-making tools. These needs have begun to impact the CIM strategies and management tactics at Lake Stevens. It has become evident to me that fewer managers are needed and that management is evolving toward a systematic and mutual responsibility of the group. It has already resulted in the elimination of one layer of management across much of the organization. Spans of control are larger in all parts of the factory. Centralized decision making is substituted with "management by objectives," and decisions are made closest to where the work is done. Morale and commitment is higher—measured by people's enthusiasm, trust, and understanding of business objectives.

The quality, cost, and cycle-time improvements achieved at the Lake Stevens plant would not be possible without TQC, automation, JIT, and participative management. Yet, as far as we have come, I am also certain that we have barely tapped the opportunity available to us. The challenge in the next few years, for all manufacturers, is to integrate these concepts into a fully functional time-based manufacturing system.

REFERENCES

Boudette, Noel E. 1989. "Networks Dismantle Old Structures." *Industry Week* 238(2) (January 16): 27–31.

Coleman, John. 1988. "Defending Demassification." *Assembly Engineering* (November): 3.

Dumaine, Brian. 1989. "How Managers Can Succeed Through Speed." *Fortune* 119(4) (February 13): 54–59.

Litchfield, P.W. 1954. *My Life As An Industrialist*. Garden City, NY: Doubleday.

Schmenner, Roger. 1988. "The Merit of Making Things Fast." *Sloan Management Review* 30(1) (Fall): 11–17.

Skinner, Wickham. 1986. "The Productivity Paradox." *Harvard Business Review* 64(4) (July/August): 55–59.

Stalk, George, Jr. 1988. "Time—The Next Source of Competitive Advantage." *Harvard Business Review* 66(4) (July/August): 41–51.

3

The Management System

The productivity of business is the responsibility of management, writes Peter Drucker. "And above all we know that productivities are created and destroyed, improved or damaged, in what we call the 'microeconomy': the individual enterprise, plant, shop, or office. Productivities are the responsibility of management" * (Drucker, 1980).

The traditional view of management is that managers carry out the function of management. Managers are people who act, direct, or control the people, resources, and functions of a business or organization. This is a narrow view, and one that is losing place to a new concept: everybody in the business has something to contribute toward management. This is the concept of participative management.

Participative management distributes responsibility, authority, and rewards across all segments of the work force. In so doing, a management system is created that requires fewer "managers" and achieves the highest levels of quality and productivity. An effective participative management system requires three structural and philosophical changes from the traditional approach: (1) information must be distributed so that decisions can be made at the level closest to where the work is done, (2) authority must be delegated so that decisions can be implemented, and (3) reward systems must be established that

---

*Excerpt from *Managing in Turbulent Times* by Peter F. Drucker. Copyright © 1980 by Peter F. Drucker. Reprinted by permission of Harper & Row Publishers, Inc.

allow people at all levels to be appropriately compensated for the contributions they make in a participative work environment.

The management system must act to utilize every resource to its fullest potential or the resource becomes a liability instead of an asset. All capital and human resources must be brought together and orchestrated to produce the most products with the highest quality at the lowest cost—hence, the definition of productivity. The management system also works to provide products and services that meet customer needs and wants in an affordable and desirable way. It must be responsive to changes in the environment to meet changing customer needs. It must also be self-correcting to ensure that deficiencies in the business are corrected. When the management system is functioning as it should, the business will be in the most competitive position possible.

A successful time-based manufacturing strategy is dependent on a participative management system. This is because time-based manufacturing also requires a broad distribution of responsibility, authority, and rewards. For example, low cycle times cannot be achieved without good quality at every process. Good quality cannot be achieved unless individuals are involved in making decisions about their process. JIT systems require scheduling decisions to be made at the local level. JIT systems require a high level of teamwork. Individuals must be motivated to make decisions on behalf of the team.

To a large extent, the present time-based manufacturing strategies evolved out of total quality control. Or at least, TQC was a necessary precursor to short cycle-time manufacturing. When TQC and time-based manufacturing are viewed together, there is an overwhelming number of contributing factors to be considered. Even a short list of management tools includes structured planning (hoshin), structured analysis—plan, do, check, act (PDCA)—statistical process control (SPC), JIT, DFM, culture, quality circles, kanban, and quality function deployment (QFD), to name only a few. A management system that addresses all of the needs of time-based manufacturing uses many different management tools.

In the years that I spent studying quality and manufacturing strategy, I noticed a pattern of interrelationships develop between the many management elements. Subsequently, I developed a conceptual model for understanding the contributions of each element of the strategy. This model was first introduced in my previous book and described as "the total management system (TMS)" (Shores, 1988).

TMS provides a framework to evaluate the management system and strategy of a business. The evaluation can be used as a means to prioritize and implement a time-based manufacturing strategy. The following paragraphs supply the background and definition of TMS.

QUALITY AND CUSTOMER SATISFACTION

The Japanese established their competitive position by focusing on quality as the basis for productivity and customer satisfaction; they require high quality from suppliers of material; they use continuous process improvement to reduce defects; they use "hoshin" to create shared plans; they use quality function deployment (QFD) to synthesize customer needs into products; and they use quality circles to improve harmony. These are the elements of TQC that the Japanese have used to improve the harmony of the factory and thereby reduce costs, improve quality, and establish a better competitive position.

In Japan, the TQC system is used as the framework for all business decisions, including market research, product definition and specification, manufacture, sales and distribution, and service and support. The TQC model most companies use is shown in Figure 3.1 (Shores, 1988). This model has "customer satisfaction" as the primary objective. The strategy is based on using quality to achieve productivity and flexibility in all processes. This TQC model is further explained in what follows.

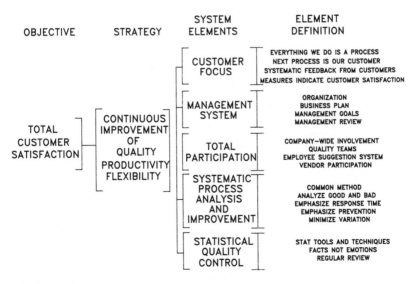

FIGURE 3.1 Japanese TQC management system

Customer Satisfaction

Customer satisfaction is the key to business success. When customers are satisfied with what they hear, see, and feel, they will come back for more. Another way to think about customer satisfaction is that customers are satisfied to the extent that their expectations and needs are met. These expectations are based on what they were told and the previous experience they had with your products and your competitor's products. See Figure 3.2.

Customers care about the product's attribute set—function, usability, performance, reliability, and supportability. Customers also care about getting the best price and best availability. To be successful then, a business must have the right product, the right price, and the right availability. The level of customer satisfaction is based on what customers hear, see, and feel about these product attributes.

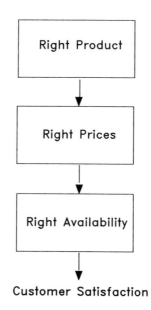

FIGURE 3.2 Customer satisfaction

Organizational Effectiveness

An organization's effectiveness is measured by its ability to provide for customer satisfaction. Therefore, an organization's effectiveness is based on its ability to (1) provide the right product in a rapidly changing environment—which requires effective planning systems, (2) provide a product at the right price—productivity, and (3) provide the needed product at the right time in the changing environment—adaptability. These organization attributes—planning, productivity, and adaptability—are the primary elements of organizational effectiveness. See Figure 3.3.

The elements of organization effectiveness represent a hierarchy of needs and serve one another. For example, effective planning systems ensure that the people in the organization are working together on the "right" things; this makes a contribution to productivity. When the processes are productive, wasted time, material, and assets are at a minimum, and the organization is capable of adapting to change in the least amount of time; therefore, productivity makes a contribution to adaptability.

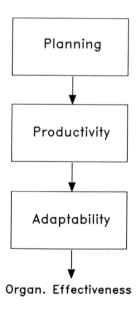

Organ. Effectiveness

FIGURE 3.3 Organization effectiveness

Total Quality Control

TQC is based on improving the quality of all the inputs to the business including information, material, labor, and equipment. When the quality of the inputs is high, processes are operated without creating defects, and, thus, defects are not passed forward to the next process; then the efficiency of each process is at its highest. Processes that do not waste time, materials, and equipment utilization are the most responsive and have the least amount of throughput time. See Figure 3.4.

TQC makes a contribution to organizational effectiveness through the horizontal relationships shown in Figure 3.5. Subsequently, organizational effectiveness makes a contribution to customer satisfaction in the same way. When these relationships are studied, beginning with quality and traced through the matrix, the link between quality, productivity, and customer satisfaction is established.

This view of the relationships between quality and customer satisfaction explains why a common focus on quality is so important to the strategy. The methods that have been developed over the years

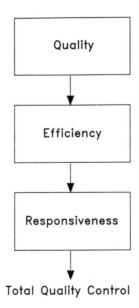

FIGURE 3.4 Total quality control

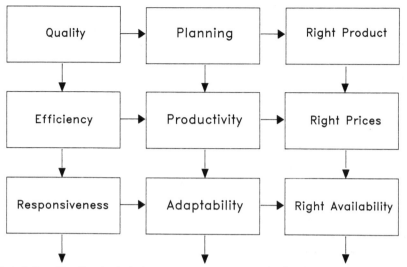

FIGURE 3.5 Quality and customer satisfaction

under the name of TQC (like SQC and PDCA) are simply a more refined approach to strengthening these relationships. Instead of relying on the vagueness of gut feel or intuition to identify process problems, employees use statistical tools. PDCA presents an organized approach to implementing solutions to these problems.

Managers and employees have many choices of the tools they use to manage their processes and business. Some of the tools are more time-effective than others and therefore have more value. In today's rapidly changing business environment, time is the most precious element in managing change. This is one of the reasons the Japanese have been so successful with TQC concepts when managing their organizations. Someone once said, "the Japanese don't manage their factories, they engineer them." The implication of this statement is that the Japanese approach business and process management much the same as engineers approach the management of systems.

In system design (like that required for the automatic pilot of a jumbo jet), the control system must react to extremely turbulent conditions. The jet's control system must sample and sense the changes in altitude and direction caused by the changing wind conditions. These changes must be analyzed to decide when and how much to compensate with the controls, and then changes must be made to the ailerons and rudder to correct the flight path. The mission of the jet will determine its size and speed, which in turn will require a commitment of resources to build it. All of the parts must be connected together mechanically and electrically to work as one system. These are the elements of system design that apply to airplanes, automatic speed controls for cars, thermostats for homes, and the control of business.

MANAGEMENT FUNCTIONS

Business organizations behave according to the same principles that govern all systems. In the "business system," the "management system" provides for the control. Through proper control, the management system ensures that the business is adaptive to the continuous changes it sees in the economic environment. By adapting to changing customer needs at a faster rate than competitors, a business assures itself of the highest level of customer satisfaction and financial success.

The five functions of management that facilitate the control and function of business are described in the following paragraphs. These functions are described here with titles that are the same as those in the Japanese TQC model shown in Figure 3.1. This is more than a coincidence. It points to the fact that the Japanese have been using these principles knowingly or unknowingly for a long time.

Management Commitment. Management commitment must be present in the form of values, investment, and individual responsibility. Management commitment provides the organization with the *cultural, physical,* and *organizational* realities of the business.

Leadership. Leadership is a process that exists in the form of visions, plans, motivation, and review of progress. Without leadership, an organization will not make consistent progress in a chosen direction. Leadership is the primary function used to *control* the pace and direction of the business.

Customer Focus. Customer focus provides for constant awareness of customer needs and attempts to meet those needs. It is the source of the *feedback* needed by the system for analysis. Without customer focus, a business will not know whose needs it is trying to satisfy or what those needs are.

Total Participation. Total participation ensures that the full genius and capability of the "people" resources are utilized. This provides for the harmonious workings of the internal parts of the organization. Without total participation, a business will depend on the genius of a few and never gain the synergy that comes from the total contribution of each person and process in the system. Total participation provides for the *synthesis* of the component parts of the system.

Systematic Analysis. Systematic analysis facilitates constant feedback, analysis, and control throughout the organization. Through systematic analysis, the deviations are *analyzed,* and the knowledge is developed to decide when and where a compensating change must be made. Without systematic analysis, an organization will be inconsistent in its response to change.

TOTAL QUALITY MANAGEMENT SYSTEM

Each of the five elements of management makes a contribution to organizational effectiveness, that is, planning, productivity, and flexibility. There are many management tools available. Some are more quality-, cost-, and time-effective than others. The matrix in Figure 3.6 summarizes some of the management methods available to modern day managers. The position in the matrix shows the function of management that the method provides by the column that it is in, and the contribution it makes to organizational effectiveness by the row it is in. Collectively, this matrix is referred to as the total quality management system (TQMS).

There are two ways to interpret the TQMS. The first is by looking at the vertical entries under each column head. Management commitment, for example, shows there are three subheads entitled Values, Investment, and Review. The entries under each subhead represent responsibilities of the management system. In subsequent chapters, detailed responsibilities are defined for management and employees for every entry in each column.

Another view of the TQMS is achieved by looking at the horizontal relationships. For example, all of the elements of management shown in the top row relate to planning. The planning row is the foundation to the entire business. This is where the culture, strategy, vision, customer needs, networking, and current situation analysis are found. All primary decisions about the direction and goals of the business are

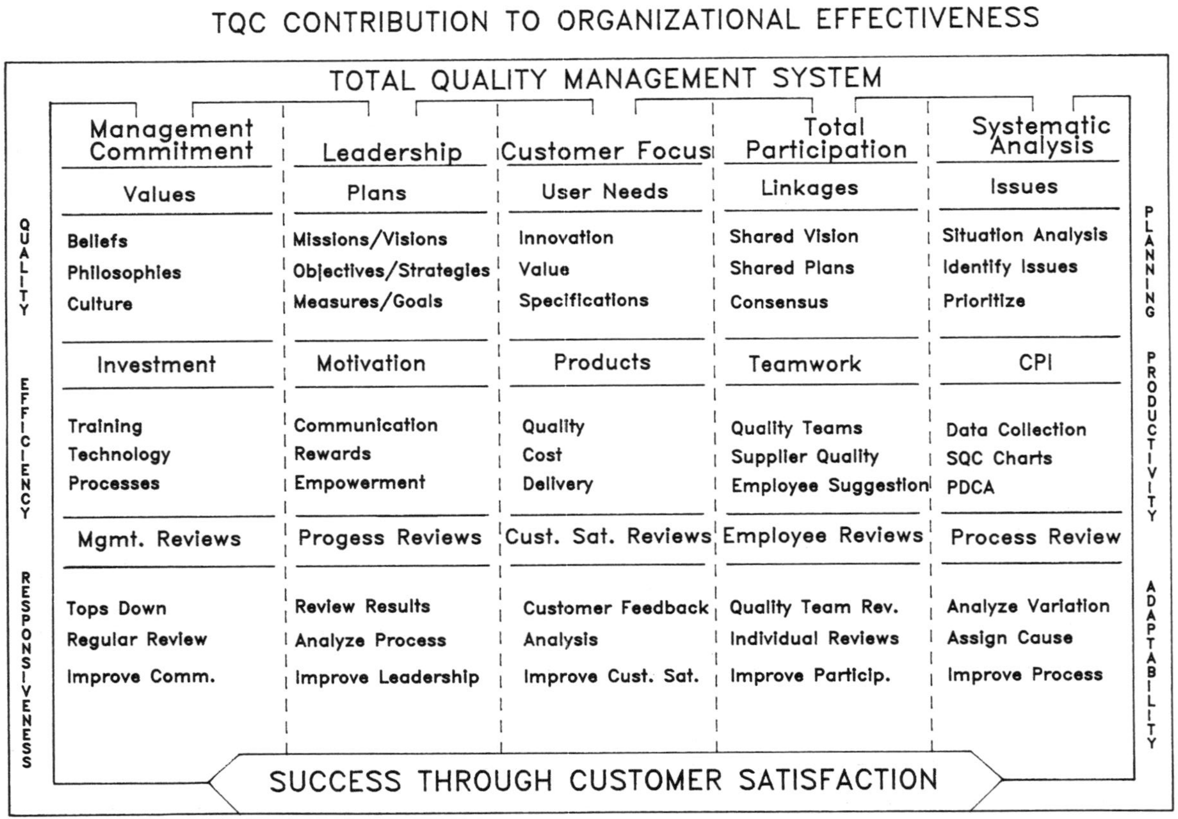

TQC CONTRIBUTION TO ORGANIZATIONAL EFFECTIVENESS

TOTAL QUALITY MANAGEMENT SYSTEM

	Management Commitment	Leadership	Customer Focus	Total Participation	Systematic Analysis	
	Values	Plans	User Needs	Linkages	Issues	PLANNING
QUALITY	Beliefs Philosophies Culture	Missions/Visions Objectives/Strategies Measures/Goals	Innovation Value Specifications	Shared Vision Shared Plans Consensus	Situation Analysis Identify Issues Prioritize	
	Investment	Motivation	Products	Teamwork	CPI	PRODUCTIVITY
EFFICIENCY	Training Technology Processes	Communication Rewards Empowerment	Quality Cost Delivery	Quality Teams Supplier Quality Employee Suggestion	Data Collection SQC Charts PDCA	
	Mgmt. Reviews	Progess Reviews	Cust. Sat. Reviews	Employee Reviews	Process Review	ADAPTABILITY
RESPONSIVENESS	Tops Down Regular Review Improve Comm.	Review Results Analyze Process Improve Leadership	Customer Feedback Analysis Improve Cust. Sat.	Quality Team Rev. Individual Reviews Improve Particip.	Analyze Variation Assign Cause Improve Process	

SUCCESS THROUGH CUSTOMER SATISFACTION

FIGURE 3.6 Total quality management system

48

made here. The planning row is the staging area from which all future action will come.

The middle row relates to the implementation. All real activity related to implementing the business plan is facilitated by the middle row. Investment, motivation, design and manufacture, teamwork, and analysis are parts of the implementation. These are the factors that govern productivity. But they do not stand alone in that respect. They work in conjunction with the planning row, which also affects productivity.

The bottom row is devoted to review. It is here where all actions related to the plans and implementation are reviewed to determine progress. It is the bottom row where secondary decisions are made related to changes in direction. Changes to commitment, leadership, products, participation, and analysis are found in the bottom row. If the bottom row is not adequately attended, the business cannot adapt to the changes in the environment and monitor its own weaknesses.

The TQMS is aimed at providing for the highest levels of quality, productivity, flexibility, responsiveness, and customer satisfaction. It is used to establish the culture and resources needed to be successful in a changing environment. It forms a participative management style on behalf of all employees. It networks all of the people and processes to function in harmony with each other and the environment. It includes the needs and thinking of the customer as the foundation of product programs. And it ensures a sound system of analysis to cope with the many changes that a business will see in the years to come.

Quality and time are the key ingredients of this strategy: quality, because no system can be productive if the material, processes, equipment, and information are defective. Time, because it is the common denominator to all productivity and responsiveness measures. The increasing competitiveness and volatility of the economic environment require businesses to conduct all actions in less time than in the past. This will continue to be true in the future.

REFERENCES

Drucker, Peter F. 1980. *Managing in Turbulent Times*. New York: Harper and Row, p. 16.

Shores, A. Richard. 1988. *Survival of the Fittest: Total Quality Control and Management Evolution*. Milwaukee, WI: Quality Press.

4

Commitment

The changing technological environment creates the need for businesses to continuously adapt to those changes. Economists tell us that technology development and its adoption are the primary contributors to productivity growth. If this is true, we are led to the conclusion that those businesses failing to keep pace with technological change will eventually lose their competitive position.

Rapid changes in technology development require faster changes in business. The ability of the business to adapt quickly is primarily dependent on its resistance to change, which can be influenced by three factors. The first is the commitment of resources to finance the changes. The second is the willingness of the employees to support and embrace the changes. The third is the organization's ability to manage change. The commitment and adherence to a set of supportive values and an adequate investment strategy therefore form the foundation of adaptability and survival in business.

VALUES

Businesses and people depend upon each other for their mutual success. Businesses need the full commitment of people's time, knowledge, and skills to achieve success measured by the growth of sales and profit. People need the success of the business and its commitment to provide employment security, pay and benefits, and career opportunities. Businesses must make these commitments and adhere to them over time to create the proper incentives for their people to help make this joint effort a mutual success.

Commitment must come from everyone in the business—top management as well as assembly-line workers. Employee commitment means that they subscribe to the basic tenets of the business and hold these things in common; in effect, they are trusting the business to do the right thing by them. Within this set of beliefs, all employees must trust that they will share in the success of the business and that the business will not treat them unfairly in the face of change or adversity.

In my research on the subject of values, I compared the traditional beliefs of ten of the most progressive and successful manufacturing companies in American industry. These beliefs were recorded by the founders or later CEOs of these businesses. Their writings are now viewed as business classics and include CEOs of IBM, NCR, Ford Motor, General Motors, Xerox, and Goodyear Tire and Rubber. I was also able to include my own experience with the Hewlett-Packard Company, a business with equal accomplishments in the competitive environment. I was impressed by the fact that these businesses have grown and changed over the years faster than their competitors. I was equally impressed by the nature of the fundamental values they hold in common.

In many ways IBM exemplifies the beliefs of the other excellent manufacturing businesses. While being one of the greatest industrial success stories of all time, IBM's beliefs are also the most focused. In his book,* *A Business and Its Beliefs,* CEO Thomas J. Watson, Jr., (1963) wrote:

> I firmly believe that any organization, in order to survive and achieve success, must have a sound set of beliefs on which it premises all its policies and actions.
>
> Next, I believe that the most important single factor in corporate success is faithful adherence to those beliefs.
>
> And finally, I believe that if an organization is to meet the challenges of a changing world, it must be prepared to change everything about itself except those beliefs as it moves through corporate life.

---

*From *A Business and Its Beliefs* by Thomas J. Watson. © 1963. Reproduced with permission of McGraw-Hill, Inc.

As a model for American industry, IBM established three values that many other business have attempted to adopt. These three fundamental beliefs are

1. Respect for the individual.
2. Give the best customer service of any company in the world.
3. Pursue all tasks with the idea that they can be accomplished in a superior fashion.

My consolidation of the "classic" values of IBM and the other businesses yielded two dominant themes: (1) respect for the individual and (2) customer satisfaction. Third and fourth values were often present but not as explicit nor as frequent as the first two. Values 3 and 4 were also given less emphasis in years past than they are beginning to receive today. They are (3) teamwork and (4) time.

Value 1: Respect for the Individual

This is clearly the most explicit and frequently stated belief. For example, P. W. Litchfield (1954) wrote of Goodyear:

> Manpower is the most important commodity with which business has to deal. That conviction had been brought home to me even in my very brief contact with administrative matters. If only a matter of self-interest, a company must respect the dignity of the individual, bringing out the best in each.

The application of this belief in these companies meant that employees are paid fair wages, provided insurance, clean and safe working conditions, educational assistance, employment security, education and training benefits, and given a voice in the company. Respect means that employees are allowed to take pride in their work. They are not asked to do things that are considered unethical. Respect for the dignity of the individual covers these things and many more.

When employees are provided with individual respect they are less afraid of change. They are inclined to have more confidence that the business will take care of them in the long run. They are not intimidated by automation or other productivity improvements because they know they will be retrained to do new jobs. Employees who share in the success of the business may be cautious toward risk but move aggressively to embrace opportunities for growth. Businesses who can engender this level of confidence in their employees will experience the least resistance to change.

Many American companies have adopted similar beliefs but, as a whole, America is still very diverse in its commitments. Respect for the individual is often missing. Sometimes the commitment to the individual is explicit in words but is missing in practice; the commitment to the individual may be absent when the pressure is on. IBM, for example, has demonstrated that their commitment to their beliefs goes beyond short-term profits. Many other companies yield to financial pressures and regularly institute unnecessary layoffs to maintain their financial image during hard times. In doing so, they fail to maintain commitment and respect for the individual.

Value 2: Customer Satisfaction

Customer satisfaction is equally prominent among these leading manufacturers. The commitment to the customer is, however, often described in different terms. IBM, for example, believes in the best customer service in the industry and being the very best in the things it does. Goodyear's belief in customer satisfaction is stated by Litchfield (1954) in two ways:

1. There is always a market for goods people have to have. But there is a potential market also for goods that are useful to people, which make their work easier or make more money for them.
2. A company is on solid ground when it is delivering value. Value means two things: goods that are useful, and a price people can pay. A company that wants to succeed must offer quality at a lower price, or higher quality at a lower price. However, as time went on, I came to put more emphasis on quality than on price. The price tag is important, but it is not the final criteria. As we said in our advertisement later on, quality will be remembered long after the price is forgotten.

The Hewlett-Packard Company describes its commitment to customer satisfaction in its statement of corporate objectives as: "providing customers with innovative products that serve real customer needs; provide products with high quality and a long lasting life; and provide friendly, courteous and economic service.'' Other companies like NCR, Xerox, and Ford have similar statements dating back many years into their history.

Today, the rest of American industry is beginning to understand the importance of quality and service to their success. In some cases, they had to learn or relearn the hard way. Ford and GM, for example, clearly started with the right focus on quality, but they lost it along the way. Consequently, the U.S. automobile industry has taken a beating from Japanese competitors. They have also paid enormous sums for recalls and legal suits over safety problems. The space shuttle explosion and the deaths of the astronauts created an awakening throughout America about the problems of negligence and quality in the aerospace industry. Numerous passenger airline crashes, believed to be caused by faulty workmanship or inadequate maintenance and inspection procedures, have brought workers and businesses to look closer at their quality standards.

Value 3: Teamwork

Teamwork has always been evident in the beliefs of most leading manufacturers. There seems to be consensus that the whole is worth more than the sum of the parts. This is clearly stated by Thomas J. Watson, Jr., Henry Ford, and in the Hewlett-Packard corporate objectives. In recent years, however, the value of teamwork is becoming even more broadly recognized. This fact is evident by the current emphasis being given to self-managed teams, participative management, quality circles, employee involvement, or quality of work life by any other name. Teamwork is becoming an important part of the new culture.

The automobile industry is one place among many where teamwork and employee involvement are on the rise. There is continued hope that automobile manufacturers will continue to receive the support of unions to jointly pursue a changing culture in the workplace (Moskal, 1989). Unions and businesses are finding that when they both share the welfare of workers as an ingredient of their mutual success, the previous conflicts begin to disappear. It is unfortunate that so many companies have taken so many years to understand the importance of employee attitudes to their success.

Value 4: Time

Time has also been an important consideration for leading manufacturers. Time to market, order cycle time, and production cycle time are examples where efforts have been successful in producing savings. One of the most noted examples of this occurred after World War I when Henry Ford was under pressure to reduce the price of the Model T. Among several cost-cutting steps, he reduced the production cycle time from 21 to 14 days. This change saved millions of dollars in inventory and carrying costs (Ford, 1922).

American industry is now finding many new ways to use time to improve productivity. In some respects, the value or belief about time is becoming part of the industrial culture. In his *Fortune* Magazine article on managing speed, Brian Dumaine cites the experiences of Domino's Pizza, whose entire business of making and delivering pizza is based on speed. Honda teaches its people to solve problems in minutes not hours. Brunswick cut layers of management and increased spending authorization for department heads from $25,000 to $250,000. This step reduced approval time, eliminates bureaucracy, and cuts down the time to develop new products (Dumaine, 1989). These changes are a fundamental part of improving the culture to ensure survival in the faster and more productive world of the future.

CORPORATE CULTURE

The commitment of a business and its people to set of philosophies is often referred to as its "corporate culture." Each business has a culture that is often based on its founder's beliefs about people, customers, and responsibilities to the world at large. At Hewlett-Packard, for example, the culture is referred to as "the HP way" and is based on the beliefs of Dave Packard and Bill Hewlett. Much of the HP way is documented in the seven corporate objectives and relate to attitudes about profit, growth, employees, customers, fields of interest, citizenship, and management. There is also an aspect of the HP way that words cannot capture; it is a feeling about mutual trust, respect, confidence, and an experience of sharing in all the good things and the bad things employees face in the environment. This culture is a key element of success for HP.

Other companies like IBM, NCR, and Goodyear are also built on strong cultures that support quality, employee needs, and customers.

This culture and the supporting beliefs are the foundation of change, productivity, and customer satisfaction. All of American industry needs to adopt similar cultures. This may be the only way to ensure a competitive position in the future.

COMMIT TO INVESTMENT

Capital investment, training and education, and process development are the primary focuses for investment commitment. If any of these are too low or out of balance with the others, the productivity of the business will suffer.

The economic environment is continuously growing and changing. Over the last 60 years, 70–80% of the nation's productivity growth has come from capital investment in new technology. This applies to every segment of the economy, from agriculture to communications. If a business expects to stay competitive, it must be willing to invest a sizable portion of its earnings back into capital investment.

Additional investment is required in process development. The business is a series of interrelated processes that functions as a pipeline for adding value to the materials and delivering the resulting product to the customers. If this pipeline is inefficient and allows wasted time and poor quality, the product will be more expensive and have less value to the customer.

A commitment to invest also means investing in human assets. Training and education are essential elements of industrial competitiveness. The people employed at every level must be able to operate the equipment and make decisions required of the processes they operate. As the industrial environment becomes more technical, the basic skills of the workers must also improve. This requires a significantly larger investment in training programs from both business and government.

Training Investment

Employees are part of a company's asset base. Some companies refer to their people as their most important asset. It is not culturally wise or practical for a business to lay off or fire people because technology has changed and the people's skills do not fit quite as well as they once did. Employees who are happy with their jobs and employers contribute much more to the organization than technical skill; teamwork, morale, responsibility, and initiative are also important contributors to productivity. A business that treats employees as important assets has a head start in human contribution. Training and educational support are part of that philosophy.

The skill and educational level of the American work force is not keeping up with manufacturing technology. Some people believe that this could be a bigger problem for American manufacturing productivity than low capital investment. As American business digs deeper into the pool of the unemployed, it is finding fewer and fewer people with the mathematical and literacy skills needed to perform efficiently in our high-tech environment. Although American business has held the reins on capital investment, government has been investing too little in education. "America, in short, has been scrimping on human capital," writes Bruce Nussbaum (1988). In his special report for *Business Week* Magazine, Nussbaum looks at many aspects of the education of America's workers compared to our competitors and the effects it has on our economy. Nussbaum writes that capital equipment investment alone won't solve America's economic ills. Just as important, if not more so, is the knowledge and skill of the workers in using the new equipment and functioning in the more technological environment.

In his book, *A Business and Its Beliefs,* Thomas J. Watson, Jr., (1963) writes: "A company must be prepared to make a commitment to internal education and retraining which increases in *geometric proportion* to the technological change the company is going through." In view of the rate of technological change business is going through today, the investment in education and retraining should be several times higher than it was 20 years ago. This, unfortunately, is not the case in public education or in business. If America Inc. is in fact "scrimping on human capital," all Americans must be willing to change their commitment to education and retraining.

A higher commitment to education and retraining requires a higher priority in government and business. Basic math skills as well as computer literacy, communication, SPC, TQC, and teamwork are important to every worker in today's manufacturing environment. Educational institutions have a tremendous challenge to train the workers of the future, but industry has a bigger challenge in training the workers of today. Aaron Bernstein reported in *Business Week* that as many as 50 million workers will have to be trained or retrained in the next 12 years. This will require industry to retrain as many as 30 million current workers and train 20 million new workers as businesses "dig deeper into the barrel of poorly educated" to find workers. The retraining will require industry to spend billions of dollars for training (Bernstein, 1988).

Other opportunities exist to better utilize people who already possess the necessary skills but are not part of the work force. This includes women who have been forced to drop out of work because of child care needs and retired employees who are victims of mandatory retirement policies. To utilize this hidden but trained work force, businesses must become more flexible. Businesses must be willing to include day care facilities, allow more part-time and flexible work hours, extend working years to fit individual abilities, and use retired employees to lead training programs.

In a *Harvard Business Review* article by Felice N. Schwartz (1989), female executives are reported as having 2½ times the turnover rate of male executives. This occurs because women leave work to have babies, do not return immediately to their jobs, and sometimes never do. Schwartz' point is that business productivity and competitiveness is seriously curtailed by attitudes and policies that do not allow women to balance their careers with their family needs. Schwartz believes that the biggest inducement to keeping women in the work force is part-time employment and flexible work schedules through "job sharing" or other creative means.

Individuals must be capable of contributing at their maximum potential in business as well as in their personal lives. Wasted assets, whether they are underutilized equipment or underutilized talent, are a drain on America's national productivity. Education and retraining are essentials to bring individual skill levels up to the level of maximum contribution. Flexible work environments are necessary to encourage people with family or other personal commitments to contribute to both family and jobs. The factory of the future must include values that encourage the respect, education, and utilization of all employees.

Capital Investment

Capital investment in new technology is believed to be the driving force behind productivity improvements. This implies that technological investment must accelerate as does the rate of technological development. In the case of human capital, Thomas J. Watson, Jr., told us that this fact is true. Increased levels of training are required to renew human resources; additional capital is required to renew capital

TABLE 4.1 Productivity Growth and Net Investment

Country	Net Fixed Investment as Percentage of GNP	Percentage Growth Rate of Productivity in Manufacturing (1971–1980)
Japan	19.5	7.4
France	12.2	4.9
West Germany	11.8	4.9
Italy	10.7	4.9
United Kingdom	8.1	2.9
United States	6.6	2.5

SOURCE: Brooks, 1986.

assets. Failure to maintain capital investment will cause a business to fall farther behind the competitors who do. On average, American industry has not invested at a high enough rate to stay competitive. Table 4.1 shows that American manufacturing lags behind the rest of the industrial world in investment and productivity.

Conversely, Japan, who is the leader in productivity gains and net investment, has the highest investment per employee (Figure 4.1) and the ratio between the United States and Japan is growing (Landau and Hatsopaulus, 1986). With regard to output per employee, Japan also surpassed the United States in 1978 and is still moving away; see Figure 4.2 (Landau and Hatsopaulus, 1986). This is an alarming trend, one that American business cannot afford to allow to continue.

Obstacles to Investment

The obstacles to investment are short-term profit motivation, bureaucracy, and management education. Top managers complain about the high cost of capital formation and the short-term profit motivations of the stock market, middle managers complain about the bureaucracy of getting major expenditures for new automation approved, and educators point to management education as the culprit. Certainly, all four are partly to blame and need to be addressed.

Short-Term Profit Motivations
Many CEOs complain that capital investment at the expense of short-term profit is harmful to the price of company stock and, therefore,

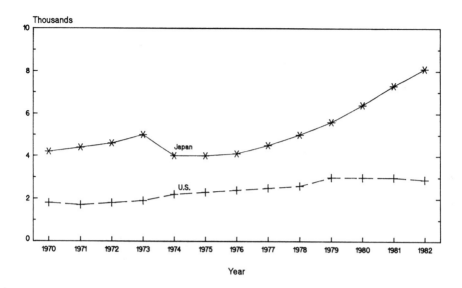

FIGURE 4.1 **Manufacturing fixed investments, per year, per employee**

SOURCE: *The Positive Sum Strategy,* © 1986, by the National Academy of Sciences, National Academy Press, Washington, DC.

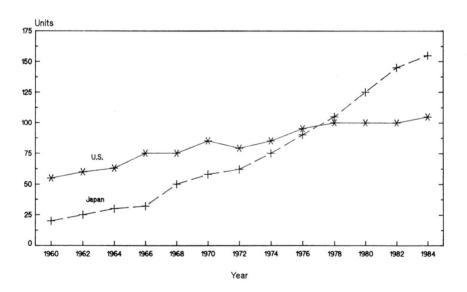

FIGURE 4.2 **Manufacturing output per labor hour**

SOURCE: *The Positive Sum Strategy,* © 1986, by the National Academy of Sciences, National Academy Press, Washington, DC.

61

unhealthy for their careers. They further argue that the cost of capital is too high because of high interest rates and depreciation requirements. Consequently, American business forsakes capital investment for alternative investments in financial markets. They keep large stores of cash invested in paper while their factories are undercapitalized and slowly headed for forced obsolescence.

Cost of Capital

In spite of the cost of capital, most of today's businesses were started by daring entrepreneurs who knew how to make a buck from their ideas. They made investments in product development, marketing, and then into capitalizing the factory. If they had not, there would not be any businesses today. Managers who believe that the cost of capital is too high to keep their businesses competitive may lack the confidence, foresight, and courage necessary to be in business. If they view short-term paper profits as more important than long-term business growth, they should get out of the business. As Henry Ford (1922) wrote: "A business that chooses to invest in financial paper at the expense of not investing in the capital required to be competitive, is not a business at all."

Assuming that the cost of capital is higher than it should be, businesses and government need to look at ways to lower it. Certainly, interest rates could be lower, but is there any chance of that happening in the short run? Realistically, how much can taxes be raised or spending cut without having a notable detrimental effect on the economy? The chicken and the egg analogy applies here. If productivity growth is the key to balancing the budget and maintaining the standard of living, then investment must come first. This would suggest that tax laws should be aimed at encouraging additional investment in process development and equipment specifically for productivity improvements.

Changing the tax laws to give specific credit to productivity investment would have a substantial effect. For example, today's laws favor investment in capacity and productivity about equally. Should this be so? I suggest that capacity expansion is somewhat self-motivating in the eye of business. When sales are up, businesses expand capacity. The benefits of productivity gains, however, have more risk associated with them, because productivity investments do not always pay off in the short run. This creates a need for additional investment incentives. Serious consideration should be given to any ideas that lead to a more productive and competitive American industry.

Cost of Noncapital

In America's quest for quality, there was an awakening about the cost of nonquality in our businesses. The same can be said about capital investment. Businesses are investing about 6.6% of revenues into fixed assets, which Table 4.1 suggests is not enough. The lost revenue and profit from low productivity is substantial. Raising the investment to increase productivity would have a short-term effect on profits, but long-term revenue and profit growth will offset this. This idea challenges the technological confidence of all managers in a business.

Financial Performance

Today's managers who do not believe that the financial markets will support short-term trade-offs for long-term investments are "just making lame excuses," writes Gary Hector (1988) in *Fortune.* Hector cites example after example of companies whose stock is valued more for its long-term growth potential than its short-term profit. Investments in new technology, new businesses, and other ventures are the basis for the confidence Wall Street places in the future of business. Short-term profits may seem to be an obstacle, but in reality, the key is in the confidence Wall Street has in the ability of a business to make those investments pay off.

How Much Is Enough?

Levels of investment vary by business, industry, and country. Some companies target capital investment to grow no faster than revenue growth, which seems to be a self-defeating strategy. How these businesses will ever get out in front of the competition, which is investing more and growing faster, remains a mystery. Michael J. Callahan writes that the U.S. semiconductor industry has annually invested over 15% of sales during the previous 5 years and expects capital investment to be over 20% of sales during the next 5 years. Much of this investment, however, is for capacity expansion and not new technology (Callahan, 1985).

Landau showed in Figure 4.2 that the Japanese investment per worker was 50% ahead of the United States in 1984. This was further substantiated by Karen Pennar (1988) in *Business Week,* who cited the investment per employee per year in Japan to be $6,500 compared to the United States at $2,600. Pennar's data also show that the Japanese growth rate of capital investment in automation is 50% ahead of the United States. This would imply that American manufacturers should be investing at a rate at least 50% higher than they are currently. Further, the accelerated rate of technological change will require that this level of investment continue to grow.

Bureaucracy
A major obstacle in large manufacturing businesses is bureaucracy. It seems the larger the business, the more bureaucratic it becomes. When new technologies are being introduced, invariably a senior executive sitting in a corporate office wants to take control of all investments. This is done in the "interest" of ensuring that all the new equipment is standard and that full use of the new equipment is achieved in one location before allowing duplication at another. In some cases, the new technology is viewed from the top as an opportunity to consolidate manufacturing processes and achieve the economies of scale the business missed in the past. Plant manufacturing managers work in frustration, waiting for their turn to come up in the procurement chain. These bureaucratic controls contribute to the reason it takes 10 years to introduce new technology instead of the 5 it should take.

A case in point is the advent of surface-mount technology (SMT), pioneered by the Japanese in the 1970s, and now a major part of their electronic manufacturing technology. In spite of the relative advantages of SMT over through-hole assembly, U.S. manufacturing lags behind Japan by 5 years. Wesley R. Iversen reported in *Electronics* Magazine that 33% of the integrated circuits (ICs) used in Japan in 1987 were surface-mount, compared to 7.6% in the United States. By 1992, the IC application of SMT is projected to rise to 60% in Japan and 35% in the United States (Iversen, 1988).

SMT is the kind of new technology that electronic manufacturers should adopt as soon as possible. It offers increased density, lower part cost, improved performance, better reliability, and lower assembly costs. In total, SMT potentially offers printed-circuit users a 30–40% production-cost advantage over through-hole technology. Iversen reported that one of the obstacles is the large investment that many U.S. manufacturers made in through-hole assembly equipment in the early 1980s.

In their efforts to squeeze additional return out of their existing through-hole capital investments, businesses limit growth in SMT investment. These companies then turn to outside suppliers for SMT assemblies and/or make limited investments in SMT equipment. The designers of new products are faced with longer lead times for prototypes, a shortage of the local process knowledge required to optimize designs for manufacturability, and a lack of confidence in their ability to be successful with the new technology. This demotivates designers from quickly adopting the new technology. When coupled with the bureaucratic controls on manufacturing investments, the time taken to adopt new technologies is doubled or tripled. The time required to adopt new technologies ends up costing American industry billions of dollars a year in lost opportunity.

Management Education

Part of the problem with top management is that they become technologically obsolete, or, in some cases, were never technologically competent. They may have come from financial backgrounds and view every investment by its short-term return on investment (ROI). In other cases, they may have once had state-of-the-art knowledge of manufacturing, but have lost their grasp of the technologies as changes have taken place. U.S. business needs more managers who have had a rounded experience in the business, much the same as Japanese managers. By rotating managers through each phase of the business, the Japanese build top managers who are technologically competent and confident in their technologies.

Wickham Skinner* (1985) of the Harvard Business school summed up the situation well:

> Those few companies which have made great gains by taking advantage of new manufacturing technologies did so by demonstrating top-level leadership and management commitment.
>
> But far more prevalent are those top managers of manufacturing firms who are neither knowledgeable nor comfortable with their industry's equipment and process technologies. This is the major educational problem: the development of technologically competent and confident top management.

Where to Invest

Investment in equipment and new technology may be applied to robotics, word processing equipment, computer-aided design/computer-aided engineering (CAD/CAE), computer integrated manufacturing (CIM), and many other manufacturing technologies. CIM is a relatively new concept that makes significant use of computers to improve the information flows from design, procurement, and assembly and test of products. CIM also is part of the time-based manufactur-

---

*Reprinted from *Education for the Manufacturing World of the Future*, © 1985, with permission from the National Academy Press, Washington, DC.

ing strategy and, therefore, an important element for every business. The Manufacturing Studies Board (1986) of the National Research Council reported the benefits five companies derived from advanced manufacturing technologies over a period of 15–20 years:

Reduction in engineering design cost:	15–20%
Reduction in overall lead time:	30–60%
Increase in product quality:	2–5 times
Increase in capability of engineering:	3–35 times
Increase in production productivity:	40–70%
Increase in capital equipment productivity:	2–3 times
Reduction in work in process:	30–60%
Reduction in personnel costs:	5–20%

Roger W. Schmenner (1988) conducted three separate surveys of many U.S. and international manufacturing businesses to understand the contributors to productivity. His study of over 265 plants shows that throughput time is the only statistically significant variable that correlates to productivity. His data revealed that a 50% reduction in throughput time impacted productivity by 2–3 percentage points. Schmenner's studies also revealed that productivity improvements were not statistically related to small companies or large companies, low-tech or high-tech companies, high-capital or low-capital companies; productivity was simply a matter of how these businesses used their resources to achieve low throughput time (Schmenner, 1988).

CIM contributes to fast throughput times in a manner similar to JIT production. By creating shared data bases and networking information sources and users together, it is possible to create JIT information flow. When accurate and easy-to-use information is available, when and where needed, the manufacturing queue time is reduced, new product introductions can occur instantaneously with a file transfer, process decisions can be made immediately, and material movement and accounting can be handled in real time. Information-processing technology is the foundation for this environment.

Information-Processing Technology
New technologies are rapidly changing the labor and skill requirements of industry. Automation, information management, and pro-

cess-improvement methodology are part of the new technology. To date, the greatest impact has been in direct labor jobs, which are being replaced by various types of automation at a rate approximately equal to 6–10% per year. For the electronics industry, where labor content is often less than 3% of the cost, direct labor could be approaching zero in 10 to 15 years.

In some industries, simplified processes and high volumes have already yielded to automation. In these factories, assembly lines and robots are busy 24 hours a day, stopping only for routine maintenance and occasional repairs. The people that are needed in these factories are higher skilled and fewer in number. They are the ones who support and maintain the sophisticated equipment, buy materials, schedule the work, and supervise those activities. In time, even these jobs will be replaced by higher levels of technology. How many people will be needed in the factory of the future depends on the growth of industry and the assimilation rate of new technologies.

Information technology has found its way into just about every segment of the economy through the use of microprocessors in automobiles, thermostat controls, toys, and in many other hidden applications. There are potentially thousands of other applications for this technology since just about every activity is a form of information processing. Ultimately, the level of technology utilization will depend only on the cost of the technology and the imagination of business.

The applications for information processing are centered around collecting diverse and complex data through a variety of media, analyzing it, making decisions about what to do, and taking some action in response to the decision. The cost is driven by the volume of the data to be processed and the time it takes to process the information to arrive at a correct decision.

Information-processing cost, therefore, yields to data simplification and cost per MIPS (what computer people call "millions of instructions per second"), which is a measure of processing speed. Advancements in computer technology contribute to steady improvements in the cost per MIPS. For example, information-processing speed has increased and the price has come down to yield a cost per MIPS improvement of several hundred times what it was 25 years ago.

This latter fact has driven the growth rate of the information economy. In electronic goods, for example, production workers decline in numbers while information workers increase in population. The information workers are charged with the responsibility to design and maintain the higher levels of automation used to replace the production workers and handle the process information.

It could be stated that both types of workers are information processors. Assembly workers are doing nothing more than using a set of information from instruction books and memory, analyzing it, making decisions about what to do, and carrying out the action. The cost per MIPS for assembly workers is unfortunately increasing, whereas the cost per MIPS for microprocessor and robotic technology is steadily decreasing. See Figure 4.3.

The technological aspect of the human being as an information processor is at a virtual standstill, whereas the cost is driven upward by training, wages, and inflation. On the other hand, microprocessors, automation, and robotic technologies continue to improve and become less expensive at very dramatic rates. Given the applications where scientists and engineers have already combined the information-processing technologies to industrial use, one could extrapolate the speed at which this would spread across the industry. The variables of this expansion rate include wage rates by industry, skill levels required, the cost-per-MIPS comparisons, and the need, which is driven by competitive pressure.

Extending information technology to new applications is limited by cost and imagination. The first applications were for number

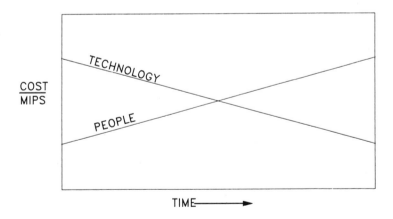

FIGURE 4.3 Cost of workers vs. cost of technology in information processing

SOURCE: *The Positive Sum Strategy,* © 1986, by the National Academy of Sciences, National Academy Press, Washington, DC.

crunching on large amounts of accounting and mathematical-type data. The output was for human consumption and decision making. As the capacity increased to handle more variables and increasing complexity, information technology was applied to the control of machinery and equipment, starting with the simplest and most labor-intensive processes first. As the cost came down and the speed went up, more complex decision making could be included. This created opportunities for extending it into artificial intelligence and manipulating the intricate movements of robots. Each year, the continuum of improvement in the cost per MIPS extends the potential applications.

Some of the processes where information technology can contribute to lower manufacturing time include procurement time, assembly and test time, engineering change time, packaging time, and production cycle time. Think of the impact on cost, quality, and flexibility if these times could be reduced by a factor of 10. Think of the impact if the time to design and sell products was reduced by a factor of 10. These thoughts may seem outrageous, but they are achievable with the proper development and application of technology.

There are many tools available for manufacturers to use to reduce cycle time and improve quality and productivity. Take, for example, a bar-code reading system. If used throughout the process to track inventory movements, it can reduce tracking errors by 70–80%. In one example, a shipping department put all of the products and accessories on a bar-code system. They also set up a pick list in the order-processing system to tell the material handlers what accessories went with which order. As the order was readied for shipment, the material handler used a computer-generated pick list to pull the accessories from stock. Each accessory was subsequently scanned as it was put into the box and compared to the code in the computer. The result: shipping errors were down by 80%. Time saved: 10–15%, because new people did not have to relearn what went into each box.

A bar-code system can also save considerable time and part errors when kanban bins are bar-code marked and are a part of the material replenishment system. There are many other technology solutions available for manufacturing today. These include defect-tracking systems, performance-characterization systems, on-line documentation, and many other information-technology solutions. These are just a few of the applications available in manufacturing. Each business is unique and each manager must find opportunities to apply the technology.

The few examples of information technology just given are a mere sample of what can be achieved in the future through adequate planning, development, and investment. Clearly, the investment in U.S. manufacturing technology has been too low in the past. Catching up with the Japanese will require a different kind of thinking on the part of management.

Commitment to Process Development

A business must also make a commitment to its processes. A process is defined as the steps required to add value to the product. In a typical situation, a process would start with some amount of material and/or information as the input and then follow a series of steps using equipment and/or labor to create the output. Each series of steps will be different, depending on the output desired and the process designer's chosen methods. In many situations, the processes are not very well designed, leading to inefficiency and poor quality. Additionally, some processes are not documented and controlled consistently, also leading to inefficiency and poor quality. Managers must be committed to providing the resources to ensure that processes are properly designed, documented, and controlled throughout the business.

Process Development

Managers usually have the choice of buying new process technology or developing it themselves. Sometimes the desired technology is not available and progress requires it to be developed internally. For example, JIT is a distribution process technology. The methods for implementing JIT are fairly well known and all that is required is management's commitment of resources to implement it. Being on the leading edge, however, often requires managers to invest in the proprietary development of processes.

Citing personal experience, I can say with emphasis that many of the gains we achieved at Hewlett-Packard were the results of proprietary process development. This sometimes included the development of special software for doing design analysis, defect tracking, and online documentation. Had we waited for these tools to be commercially available, we would still be waiting. As it is, we saved millions of dollars in productivity gains that are not yet available to other businesses.

Developing processes requires a commitment and confidence on the part of management that it can achieve a payback on these investments in a reasonable period of time. The payback is never certain, however, and will always entail some risk. Managers who are timid or not technologically confident or competent will find themselves unwilling to take these risks. They will subsequently find themselves running in the middle of the pack.

Process Implementation

In Japan, businesses not only invest more per employee, but they get more output per hour worked, as Figures 4.1 and 4.2 showed. Another factor is that they get more output per dollar invested. Process improvement through the implementation of JIT, reduced setup times, and reduced learning curves has a direct effect on this result. JIT allows for less space, which reduces the facility investment and improves overall productivity. Reduced setup and learning curve times, which must come with reduced cycle time, improves equipment utilization. These factors combine to improve the output per investment dollar.

There are many other processes in a business where these principles have equal value. To maximize the output per dollar invested in capital, every activity in the business must be viewed as a process. This is referred to as "process-focused management." Typing a memo, sweeping the floor, and designing a computer are different types of processes, but processes nonetheless. Other processes include assembling a car, a computer, or airplane. Simple or complicated, processes have three things in common: (1) an input of material, information, and labor, (2) a prescribed series of steps or actions to get the work done, and (3) an output or end product. Thinking of everything as a process with these three common factors helps to bring the common aspects of processes into focus and also helps to simplify the process-improvement task.

1. Input of material, information, and labor: requires a definition of the material specifications and the skills required. This determines whether or not it must be done by machinery, robots, or people.
2. Series of steps or actions to do the work: the design information and process steps based on the anticipated equipment and skill level. This establishes the level and type of documentation required by the operators.
3. An output or end product: defines the quality and performance expectations for the product.

When everything is viewed as a process and everyone is committed to improving all of the processes, then the business is committed to continuously improving quality, cycle time, productivity, and customer satisfaction.

REVIEW

Commitment to a set of values and investment strategy implies that the commitment will be met over a period of time. Along the way, circumstances will arise that suggest that the wrong strategy is being pursued or the economic environment will create the temptation to scrimp and cut corners. Reviews are necessary at every level in the business to ensure that the values and investments are maintained at a level adequate to sustain desired performance.

The reviews require a commitment of time by all people in the organization. These reviews must be appropriately broad at the top and become narrower in focus as the employees review their own work.

A review, in itself, can become bureaucratic if overdone. It may make employees and managers feel as if they are not trusted to do their work. Further, managers may spend endless hours of staff and administrative-support time preparing for formal presentations. The CEO of Brunswick was reported by Brian Dumaine as asking for a weekly one-page summary of the progress and needs of his staff. Each manager must study his or her own conscience to establish the proper balance of time for doing and reviewing (Dumaine, 1989).

Top–Down Review

Japanese management systems that utilize TQC contain a presidential review process. This review is intended to assure that progress is continuously being made throughout the company toward implementation of TQC. The Japanese companies that use these reviews have the same good reasons for using this process. Each felt that the employees needed a common goal. Also, having a presidential review allowed them the opportunity to provide recognition for individual achievement and constructive feedback to help those who needed it. The goals of the review process might be stated as follows:

- To ensure that performance plans and goals are achieved using a common method such as TQC.
- To ensure that the performance of the process is being controlled with the aid of SQC tools and techniques.
- To reveal the strengths and weaknesses of the system for improving the quality of the product or service.
- To ensure that performance measures and goals are realistically set to reflect current problems and issues.

In structuring their organizations for reviews by the president of the company, the Japanese have demonstrated that each strategy in the hierarchy has significance to the mission of the business. They use very detailed planning systems to ensure that every goal and measure is tied to the next level in the hierarchy. People in the organization are not allowed to waste resources on issues that do not specifically relate to the mission of the business and the objectives. The presidential reviews established are specifically designed to provide the visibility and recognition required to ensure that all resources are focused on the mission of the business.

Some aspects of management commitment cannot be measured adequately by the normal review process. Such is the case with the maintenance of the corporate values and culture. Employee attitudes toward management may be waning and subsequently morale is not what it should be. Employee surveys are very good at determining employee attitudes. Surveys can be used to see if the values are in fact being used and followed or simply written and talked about. When conducted regularly, surveys can be used to benchmark employee attitudes and work for continuous improvement.

These types of surveys also require a commitment of management, both in terms of the cost of conducting the surveys and the necessary follow-up, analysis, and action to make improvements. The improvement process can be thought of in terms of the standard USA–PDCA process (see Chapter 9). Understand the issues, select the biggest issue, analyze the cause, plan an action, do the plan, check the results, and act on the results. This is management TQC.

Regular Reviews

Reviews are generally conducted at regular intervals, determined by the times they are expected to take to see the results. The times will be longer at the top (possibly months), monthly by middle management, and maybe weekly, daily, or hourly for individual workers. The objective of these reviews is not to "watchdog" employees, but to help identify flaws in the strategy and the investment. Specific actions coming from these reviews should not be punitive to individuals, rather, they should result in changes to strategy and investment.

Conducting regular reviews at all levels requires a commitment of time. Very often, managers and individuals do not commit the time to do these reviews; the consequences are that the motivation to continue investing and improving the process disappears. People simply do not feel as though they are doing anything important when reviews are not done. Subsequently, the organization does not make the improvements required to remain competitive, and it begins a long decline to obsolescence.

Improve Commitment

When the reviews are conducted at appropriate time intervals for all levels, and the results are reflected in tighter adherence to values and better quality investments, the business will then be in a position to continuously improve competitive performance.

REFERENCES

Bernstein, Aaron. 1988. "Where The Jobs Are Is Where The Skills Aren't." *Business Week* 3070 (September 19): 104–108.

Brooks, Harvey. 1986. "National Science Policy and Technological Innovation." In *The Positive Sum Strategy*, edited by Ralph Landau and Nathan Rosenberg. Washington, DC: National Academy Press.

Callahan, Michael J. 1985. "Manufacturing Issues in the Semiconductor Industry." In *Education for the Manufacturing World of the Future*. Washington, DC: National Academy Press.

Dumaine, Brian. 1989. "How Managers Can Succeed Through Speed." *Fortune* 119(4) (February 12): 54–59.

Ford, Henry. 1922. *My Life and Work*. Garden City, NY: Doubleday.

Hector, Gary. 1988. "Yes, You Can Manage Long Term." *Fortune* 118(12) (November 21): 64–76.

Iversen, Wesley R. 1988. "Surface-Mount Technology Catches on With U.S. Equipment Makers—Finally." *Electronics* 61(18) (December): 116–119.

Landau, R., and N. Hatsopaulos. 1986. "Capital Formation in the United States and Japan." In *The Positive Sum Strategy*. Washington, DC: National Academy Press.

Litchfield, P. W. 1954. *My Life As An Industrialist*. Garden City, NY: Doubleday.

Manufacturing Studies Board, National Research Council. 1986. *Toward a New Era In U.S. Manufacturing*. Washington, DC: National Academy Press.

Moskal, Brian S. 1989. "Quality of Life in the Factory: How Far Have We Come?" *Industry Week* 238(2) (January 16): 12–16.

Nussbaum, Bruce. 1988. "Needed: Human Capital." *Business Week* 3070 (September 19): 100–103.

Pennar, Karen. 1988. "The Productivity Paradox." *Business Week* 3055 (June 6): 100–103.

Skinner, Wickham. 1985. "Challenges to Be Met." In *Education For the Manufacturing World of the Future*. Washington, DC: National Academy Press.

Schmenner, Roger. 1988. "The Merit of Making Things Fast," *Sloan Management Review* 30(1) (July/August): 11–17.

Schwartz, Felice N. 1989. "Management Women and the New Facts of Life." *Harvard Business Review* 89(1) (January–February): 65–76.

Watson, Thomas J., Jr. 1963. *A Business and Its Beliefs*. New York: McGraw-Hill.

5

Leadership

The direction and pace of a business is established by the effectiveness of the leadership process. The business organization is guided by its perceptions of the changes and opportunities in its environment, much like a car is driven down a road. The driver watches for turns and other vehicles, steering to stay on the road and avoid collisions. In a business, the leaders must be cognizant of the changes in technology, economic environment, and competition. Changes that go unrecognized in these areas can spell danger for the business.

The rate of progress of a business is similarly affected. Too many changes in the direction of the business slow its forward progress; too little horsepower (motivation) is also detrimental to progress. In the race to move forward with quality, cost, and the delivery of innovative products, consistent progress at the fastest pace is essential for success. Providing the direction and motivation for the business is the function of leadership.

Effective leaders must possess the ability to perceive subtle changes in the structure of business, world power, politics, or the economy. They must also have the ability to move resolutely toward creating the enthusiastic and united support of many people. Sometimes leaders must convince others that hidden perils threaten to devastate a way of life. Other times, they must seize the moment and take advantage of fortuitous circumstances that come along once in a lifetime. In either event, the leaders, be they in positions of great political power, revolutionaries, or business, must be able to influence change in large numbers of people's thinking and behavior.

The traits of a good leader include first and foremost an appreciation for the past, present, and future. The old adage still applies that: "We can't get to where we want to go if we don't know where we are." Good leaders recognize that the here and now holds the key to future events by the connection between present knowledge, learning, and future outcomes. They realize that continuous learning causes a continuum of technological and economic change over time. Some of the changes that will take place in the future are cyclic and, therefore, predictable; others are built on the achievements of the past and are, to some extent, also predictable; very few changes are random and unconnected to the past.

Good leaders also realize that the continuum of technological growth requires that the behavioral patterns of the people and institutions must also change. The challenge to leaders is to use their perceptual power in a visionary way to communicate the needs and opportunities of the future. Leaders then use every persuasive means in their power to create the supporting human motivation.

A leader's success is marked by the group or organization achieving a more harmonious relationship within its environment. Success is evident when both the needs of the organization and the people are served equally well. To the extent that the people's needs are best served by the success of the organization, it is the people who ultimately and always benefit or suffer from the quality of leadership.

The difference between "good" leadership and "bad" leadership is frequently associated with personality. Charisma, for example, is a personality trait that some scholars have included in the definition of good leadership. Peter Drucker (one of the more influential modern thinkers on management and leadership) disagrees with this thinking. In an article for the *Wall Street Journal,* "Leadership: More Doing Than Dash," Drucker credits hard work and hard thinking as applied to creating missions, goals, priorities, and standards; an acceptance of responsibility for the outcome; and a high level of integrity to build trust among the people who are to follow. Drucker dismisses the value of charisma, stating that it can contribute to a leader's undoing by bringing about a belief in that individual's own infallibility. Leaders then become inflexible and unable to see the errors in their ways (Drucker, 1988). Drucker insists that there is no common set of desirable "leadership qualities" or "leadership personalities"; there is only hard work and a *means* to achieving the performance of the organization.

When personality attributes are removed from leadership, what remains is a leadership process—the means. As Peter Drucker stated, good leadership is the result of hard work, establishing clear objectives and goals, setting down standards of implementation, and creating motivation. When review is added to this sequence of leadership functions, a process is established that allows for continuous improvement toward the objectives. This is the "leadership process." There are three primary elements to the leadership process: planning, motivation, and review.

Modern businesses are beginning to recognize that leadership cannot be limited to the hands of a few managers. On the contrary, they are learning that the needs of the business are best served when leadership is distributed among many employees. Businesses benefit by having more coherent goals, higher levels of motivation, and fewer management people. In his book, *The Leadership Factor,* John Kotter refers to this type of leadership as "leadership with a small (l)," as opposed to the "larger than life capital (L) Leaders" (Kotter, 1988). Businesses need more (or need to provide the motivation for) people to see their roles as providing leadership from every corner of the business.

Business leadership begins with an entrepreneur's perception of changes in the environment. These changes may be seen as problems with which to deal or opportunities. Effective leaders see the opportunities and have enough knowledge to create a vision for taking advantage of them. The vision is in the form of a product or service. In a start-up business, the perceptions and visions are essential and fundamental roles of the entrepreneur, who must then follow up with a commitment of resources and provide the motivation to make them happen. This is what John Kotter refers to as leadership with a capital "L."

As a business grows and takes on larger numbers of people and established products, it requires more organizational leadership. This form of leadership comes from many people of the organization and is the responsibility of more than one person. The business has many people in contact with the outside markets. These markets are rapidly changing and presenting new opportunities. The internal parts of the business and its technologies are also changing. The internal and external changes provide opportunities for improvement in the products and processes. In a large and dynamic business environment, there are many people involved in interpreting these changes; therefore, a system must be in place to collect and prioritize the many perceptions and create the shared vision or visions of the business. This is the role of organizational leadership.

Creating organizational leadership means establishing a system for sharing leadership roles. This is done through shared planning, shared motivations, shared reviews, and shared decision making. Shared leadership means that all employees from top to bottom share in the knowledge, development, and ownership of the elements of leadership.

PLANNING

The planning process is the tool used to establish a common set of goals and strategies for the people in the business to work toward. An effective planning system keeps everyone working on the "right things" and minimizes the wasted effort and time that occur from going down dead-end paths. An effective planning system also keeps everyone working on the "critical few" strategic elements of success. This prevents too broad of a dispersal of the resources. Given a finite set of resources, the business can choose to move slowly on a broad front or move forward quickly with a highly focused thrust. The planning process enables the business to create the focus that it chooses.

Purpose

To start a business, the leaders must be cognizant of the environment, its needs, and its opportunities. With that knowledge, the business must shape its "purpose." The purpose is the reason the business exists. It usually makes a statement about which customer needs the business intends to satisfy, and defines the boundaries for its field of interest. When everyone in the business stands to benefit from fulfilling the purpose, it can be highly motivating. This is true, however, only if the values of the business establish equitable sharing of the success.

Each function and process also has a purpose that should be aimed at helping the business meet its goals. After the purpose is developed at the top, it should also be developed at each level in the organization. This ties each department to one another to close the "cracks" and eliminate redundancy. As each purpose is developed, all persons should achieve consensus between their managers and peers. All should be knowledgeable of their department's purpose and the purpose of their job.

Example

Purpose of Manufacturing Manager

Provide labor, material, equipment, and manufacturing processes that provide the XYZ business a competitive advantage in the area of quality, cost, and delivery.

Purpose of Process Engineering Manager

Provide processes that give the business a competitive advantage in quality, cost, and delivery.

Purpose of Materials Manager

Procure materials and make them available to the production processes to give this business a competitive advantage in quality, cost, and delivery flexibility.

Purpose of Production Manager

Provide labor and equipment to assemble and test processes to give the business a competitive advantage in quality, cost, and delivery.

Purpose of Assembly Supervisor

Provide labor and equipment to all assembly processes, etc.

Purpose of Widget Assembly Worker

Provide labor to the widget assembly process to give this company a competitive advantage of quality, cost, and delivery.

Vision

After the purpose is established, the business needs a vision of how it is going to fulfill the purpose. The vision is usually specific about the type of products or services that the business will deliver to fulfill the need. For example, Henry Ford's vision was of an automobile (horseless carriage at the time) powered by an internal-combustion engine. The vision was to satisfy a need—fast, efficient, horseless transportation. Later, after that vision was accomplished, the vision became one of mass-produced automobiles, and so on. Henry Ford perceived a need for lower-cost, faster transportation. The need provided an opportunity to sell many cars (Ford, 1922).

The vision should be founded on realistic ability, technology, and opportunity. When it is, it becomes a motivating tool for the rest of the organization. It is the leader's role to build this vision and communicate it in a motivating way. If the vision is not understandable or considered realistic, it is doubtful that it will motivate many people to work toward its achievement. People can be hired to do what they are told, but the power of human motivation that comes from a shared vision will be absent.

Every employee needs to understand the company's vision. In some cases, they may develop their own visions that are consistent with that of the business. For example, the vision may be stated as: "develop and sell a laptop computer for under $500 that provides basic word processing capability, 10-year life, and makes 10% net profit, by November 1989." The design manager may have a vision for an integrated circuit that costs $5 and does everything needed. The product designer may have a vision for a two-step assembly process, etc. Collectively, these visions should meet the requirements for making the vision of the business a reality.

Vision for Time-Based Manufacturing
A time-based manufacturing strategy also requires a vision. This vision should address the elements of speed, information, people, and processes. The vision that I have for the factory of the future is stated as follows:

> A manufacturing process where inventory queues are not necessary to meet product delivery flexibility requirements; people and processes share information and make decisions in real time without management intervention; and all employees share a sense of pride and ownership for the success of the business in meeting quality, cost, and delivery goals.

This vision captures the essence of the strategy defined for XYZ Manufacturing in Chapter 2. As a strategic manufacturing vision, parts of it can become the responsibility of different departments in manufacturing. At the same time, it brings focus to the various departments to help them work together on achieving mutual goals.

Measures and Goals

The term "measure" is defined as a description of the information that will be evaluated to measure progress. For example, "defect rate" is a measure of quality. The "goal" is defined as the actual number the measure is planned to achieve. Example:

Measure	*Goal*
Defect rate in PPMs	100 PPMs

All people and processes should have a set of measures and goals for their strategies and daily work tasks. Care should be taken, however, to ensure that all measures and goals are under the control of the employees doing the measuring. Employees need to be responsible and accountable for their measures and goals in such a way that if progress is not being made, they can do something about it.

The measures and goals selected for the time-based manufacturing function should relate directly to the vision. The first question asked should be: How is the progress toward making the vision a reality going to be measured? In the previous example, progress on the inventory queues can be measured by the levels of raw parts inventory, work in process, and finished goods inventory. Progress on information sharing and decision making, however, is more difficult to measure and may require more indirect measures. The number of supervisors and managers may be an indicator of the amount of indirect support required to handle the process information and make decisions. Delivery flexibility is also difficult to measure and may be equally served by alternative measures. Cycle time, for example, can be an indicator of the delivery flexibility available.

In a manufacturing organization, there are three fundamental measures of success and strategic importance. These measures are quality, cost, and delivery flexibility. Every department in manufacturing has a set of measures that it shares or a subset of these measures. Table 5.1 is an example of the measures for a fictitious manufacturing operation.

TABLE 5.1 Manufacturing Measures Matrix

MANUFACTURING	MATERIALS	PRINTED CIRCUIT	ASSY, TEST & DISTRIBUTION	INFO. PROCESS SUPPORT	MFG. ENG.	FACILITIES
QUALITY						
• Customer Failure Rate	Mfg. Part Failures	Part Defect Rate at Inst. Test	Early Fail. Rate Ship. Quality	# System Failures	Customer Failure Rate	Cust. Satisfaction Workers Comp. # Service Interruptions
COST						
• Prod. Cost % GMP	Exp. vs Tgt. % GMP / Tot. Cost/Comp.	Exp. vs. Tgt. % GMP / Tot. Cost/Comp.	Exp. vs./Tgt % GMP / Tot. Cost/Comp.	Exp. vs. Tgt. % GMP / Tot. Cost/Comp.	Exp. vs. Tgt. % GMP / Tot. Cost/Comp.	Occ. Rate vs. Tgt. / Mfg. sq. ft. vs. tgt. / Occ. $ % GMP
FLEXIBILITY						
• Prod. Cycle Time • Prod. Cycle Time • Proto Type TAT • Published Avail. • % On-Time Ship. • Revenue vs. Plan	Supplier Lead Times Receiving Cycle Time Inventory Weeks Line Stop/Duration Total # Comps	P.C. Cycle Time P.C. Service Level Prototype TAT	Prod. Cycle Time Availability On-Time Ship. Revenue vs. Plan	MRP Cycle Time # Service Requests S.R. Cycle Time	2% Preferred Parts	On-Time W/O % Interrupt Duration

Strategies and Tactics

Strategies are the specific actions that are to be taken to achieve the goals. The strategies may relate to the allocation of resources, the hiring of people, the purchasing of equipment, the reassignment of people, and alignment of the organization. Generally, the top managers develop strategies that relate to the goals at the top. The goals become more focused further down in the organization as do the strategies. For example, the strategy at the assembly level might be to assemble a complete product at one time, by one employee. Or it could be to set up an assembly line and each employee adds a part in a progressive manner. Each of these strategies has different implications for how equipment and people are deployed. Sometimes the substrategies, such as equipment and people deployment, are referred to as tactics.

The summary of the strategy for the time-based manufacturing function should take a form similar to the matrix in Table 5.1. The strategy matrix, however, has the major strategic areas of focus as the columns instead of the departments. See Table 5.2.

The strategic areas across the top are elements of the vision called out before. There should be specific strategies for each area of focus and each measure. The details of these strategies vary for each business. Each strategy should make a contribution toward a measure and goal and may vary by department. In other words, production may have a different focus than manufacturing engineering on the strategy to influence quality; for example, production is working on process defects while manufacturing engineering is working on design defects.

This matrix is useful for looking at individual departments and their contribution to the measures. It is also useful for identifying the generic tools that will be used to attack each goal. In Table 5.2, there are several different acronyms in each box. These acronyms identify the management strategies to impact the measure and goals. The definitions for each of the acronyms on the summary follow:

TQC: Total Quality Control
MBO: Management by Objectives (Participative Management Plan)
TNG: Training Plan
CIM: Complete Information Management/Computer Information Management

TABLE 5.2 Time-Based Manufacturing Strategy Summary

	PEOPLE	PROCESS	MANUFACTURABILITY	INFORMATION
Quality • Customer Failure Rate	MBO TQC CIM TNG	PTP TQC CIM TNG	CIM DRP MCP TNG	CIM TNG
Cost • Prod. Cost % GMP • Labor • Material • Overhead	MBO TQC CIM TNG	PTP CIM TQC TNG	CIM MCP RDP TNG	CIM TNG

Flexiblity
• Material Through Put
  Days
• Production Through Put
• Capacity +/-/Month
• % On-Time vs. Org.
  Ack.
• Published Avail.
	HRP
	MBO
	CIM
	TNG
	FME
		PTP
		CIM
		TNG
		MMP
		FME
			CIM
			MMP
			TNG
			FME
				CIM
				TNG

HRP:	Human Resource Plan
FME:	Flexible Manufacturing Environment
PTP:	Process Technology Plan
MMP:	Material Management Plan
DRP:	Design Rules Plan (Design for Manufacturability)
MCP:	Material Cost Plan
RDP:	Redesign Plan

Hoshin

The Japanese achieve deployment through a planning system called *hoshin* (Shores, 1988). Hoshin means "policy," or "policy deployment." Hoshin is used to deploy the business policy—objectives and strategy—through the organization. Hoshin is a very hierarchical system and starts with the president of the company. The president states his critical few hoshins, which include an objective statement, a measure, a goal, a strategy, and an owner of the strategy (who's accountable). The president's hoshins are normally limited to the two or three most critical success factors for the business. When consensus for the hoshins is achieved with the next level down, they begin to develop their own hoshins. This process is repeated throughout the business until the plans are complete for every operation.

The hoshin plan is heavily documented at each level. For example, it uses tables with numbers for each objective and associated strategy. These numbers start with the president as hoshin numbers 1, 2, and 3, and progress to the next level as hoshin numbers 1.1, 2.1, and 3.1. In this manner, connectivity of the organizational network is tracked through the system. The hoshin is then used as the basis for all future progress reviews.

MOTIVATION

When the purpose, vision, and strategies have been adequately defined and shared, much of the task of motivation has been done. However, employees have other needs and reasons for working, and these are the elements of motivation that are now addressed. Communication, reward systems, and individual empowerment to act are the elements of motivation. Recall that at any given time, some percentage of the population may not understand the purpose, vision, and strategies. This presents the need for effective communication skills and systems.

Communication

Change, change, change seems to be a way of life and it is accelerating more all the time. The dynamics of the business environment and our own desire to implement a more productive factory cause even more change. The people in the organization need to be part of the change and kept informed of everything that is going on or morale problems will creep into the business and destroy the expected gains.

Some organizations have outstanding plans and planning systems, but nobody knows anything about them except top management. When employees get the feeling that they do not know what is going on around them, they begin to lose confidence in management. Employees complain, but management does not hear them; rumors get started (usually bad ones) and management does not address them; employees start quitting and management wonders why. Some of the biggest morale problems result from poor communications between employees and management and management and employees. This is a two-way communication process that must always be kept open through multiple paths.

The lack of good communications doesn't just hurt the morale of people working on the production line; it has equal effect on the ranks of professional and middle managers. This fact has been evident most recently among middle managers whose numbers are being squeezed by self-managed teams, better information systems, and increasing spans of control. Many managers who are directly affected by downsizing in their organizations feel that they are being ignored, and top management isn't communicating squarely with them. They "learn more about their future in the cafeteria than they do from their bosses," observes John A. Byrne (1988).

The human intellect, although very powerful when directed and focused on a task, also has a creative imagination. When the source of continuous and solid information flowing into it is interrupted, the mind begins to create its own idea of what is happening. This process is aided by the rumors emanating from the imaginations of other people in the environment. Consequently, the attitudes, morale, and behavior all suffer, and the organization's productivity suffers.

Continuous open and honest communication is the only way to avert or improve this situation. Thomas J. Watson, Jr., (1963) described the importance of maintaining communication channels this way: "communication must be up and down and side to side with multiple paths. Some paths don't always work."

Multiple communication channels are provided through quality teams, coffee talks, management by walking around (MBWA), regularly scheduled department meetings, newsletters, and an open door policy. Some of these communication devices best serve the transmission of information; others provide valuable insight into the concerns of employees, which could be the source of morale and productivity problems. MBWA and the open door policy, for example, are the best tools for getting input from individuals.

MBWA is typically a one-on-one exchange of information. A top-level manager who takes the time to walk out onto the production floor and hold a serious conversation with an employee can learn more about the attitudes within the organization in one afternoon than would be possible in a month of staff meetings. The concept of MBWA may be misleading for some, because the manager is not out there to redirect the employees from their assigned tasks. The purpose of the visit is to communicate openly and honestly, attempting to identify morale problems, rumors, and other, sometimes subtle, inhibitors to productivity. After a few visits with different employees, common patterns of concern may start to be visible. These concerns then become the priorities for management on which to act and discuss at future group communication sessions.

An open door policy encourages people to come and ask higher-level managers questions about things they do not understand as communicated by their boss. It also builds trust and mutual respect and lowers the barriers that exist between employees and managers. Closed doors to offices discourage this form of communication. Another barrier to this form of communication is the necktie. Many people believe that neckties worn only by managers and professionals suggest a form of class distinction and cause a reluctance to associate. The biggest barrier to using the effectiveness of the open door policy is recrimination. If an employee is singled out by the boss and accused of being a big mouth or otherwise intimidated in any way, the employee will not walk through the "open door."

Rewards

The reward structure is one of the most important aspects of the leadership process. Employees need to understand that they will be rewarded for their achievements, not penalized. Consider the following scenario: A group of employees is told that they must produce quality products and are paid on a merit-pay program. The only reliable information available on the process comes from accounting, and that information is normally based on *quantity* produced per day and time standards. Each employee comes to the job with different skills and energy; therefore, they have different abilities to produce high quality at an acceptable speed. If the pay of slower employees is penalized because of their speed, they will cut corners on quality. Quality then becomes the primary variable in the process. Unfortunately, the quality of the work is not adequately measured, therefore, much of the poor quality goes unnoticed and is passed on to the customer.

As this scenario depicts, the measure and reward of employees determine how they will act. In today's business environment, the reward systems are receiving a lot of scrutiny and, in some cases, resulting in a lot of change. Profit sharing, merit pay, or variable pay based on performance, lump-sum bonuses, and pay for knowledge are receiving more attention these days, even among unions. Nancy J. Perry (1988) reported that 75% of all pay systems now use some form of nontraditional pay (compensation other than fixed hourly or salary pay). Furthermore, 80% of these pay plans were enacted in the last five years.

There are many pitfalls to these new pay plans that must be considered and planned for when setting up incentive pay programs. Some of the problems relate to inadequate measures for quality, incomplete business plans, and measures that do not relate to performance. These problems are particularly relevant to the time-based manufacturing facility because the primary measures are changing. Addressing these problems in the context of traditional organizations will be a mistake.

In a time-based organization, JIT production is a way of life. In a pull JIT system, the operator pulls the next work forward when "finished" with the first. The operator is in a position to make that decision and must treat quality as the primary measure of importance. If quality is not first, then defects will be passed on to the next person, who will be delayed in rework. Ultimately, this JIT production process will come to a halt. Also evident in JIT environments and team-oriented structures is the loss of performance measures for individual contributors. Speed must be sacrificed for quality; the production process runs at a pace established by the effectiveness of the team, not the individual; increasing spans of control cause supervisors to lose the ability to objectively evaluate employees on speed and quality. The answers to these potential motivational problems are different for the time-based manufacturing facility than they were for the traditional factory. Building a time-based factory also means restructuring the reward system to establish the correct motivations.

Empower Employees

The highest level of motivation is achieved when employees feel that they are an equal part of the organization. If people are allowed to make decisions that affect their destiny and well-being, they are inclined to be very creative and highly productive in the task of meeting their goals. Managers empower people to reach new heights of achievement when they create an environment where these things are possible. People need to feel that it is okay to make decisions concerning their process; they need to feel that it is okay to tell managers when things are not going right; they need to feel that it is okay to stop the production line if quality has gone sour. People are empowered by a sense of belonging, team spirit, commitment, and trust.

One of the seven keys of business leadership reported by Kenneth Labich (1988) is: "Trust Your Subordinates. You can't expect them to go all-out for you if they think you don't trust them." In his special report, Labich cites the experiences of CEOs and managers who have successfully harnessed the power of their people through delegation, commitment, support, teamwork, and trust. Giving employees the freedom to make decisions and act on them is the essence of empowerment.

In a time-based manufacturing environment, all employees must be empowered to make decisions and act as required to keep their processes functioning within the framework of the business strategy. This means that managers must look long and hard at why they insist on keeping full control of the purse strings; they must question their own need to review every decision; they must prevent themselves from being drawn into levels of detail that go beyond their need to understand. Only when managers have separated themselves from the emotional attachment to this power will the organization be ready to move into the future. Where employees do not appear to have the skills to decide and act in this type of environment, management's tasks should be to provide them with the "commitment" of training.

PROGRESS REVIEWS

Reviews set the stage for understanding the condition of the leadership process. An effective review brings forth the relevant information, ensures that deviations from expectations are analyzed, and that appropriate resources are assigned to respond to problems. Every employee should be a part of the review process: production workers review their progress toward their process measures; supervisors review their groups' progress; and so forth. If changes to plans are made, reviews are necessary to see if the desired effect is taking place; if a new communication plan is initiated, reviews are necessary to see if morale is progressing. When a new plan is started, reviews are necessary to see if the motivations are forthcoming.

Review Results

All persons in the organization must be constantly aware of the progress toward their goals as defined by the measures of performance. The results may appear to be going well, but left unattended, subtle warning signs may go unheeded. This is analogous to the jumbo jet flying under automatic pilot. If the entire crew were to go to sleep, a piece of equipment could malfunction, causing the aircraft to lose altitude and crash into a mountaintop. Even if something in the aircraft does not go wrong, it may be necessary to change course; possibly, the airport of destination is closed due to fog conditions. For these reasons, constant vigilance is always necessary.

In a business, desired changes never occur as planned, and the environment is continuously changing unexpectedly. This requires that the business leaders constantly monitor their performance, looking for unexplained deviations. When leader is spelled with a little (l), it means that all employees are part of the management team, giving them a role to play in monitoring performance. When something goes wrong, it is neither productive nor wise to say, "that's not my job."

Review the Leadership Process

The review of the leadership process relates to the tools used to practice leadership. The planning model, for example, may be efficient or inefficient. If it is efficient, some people may not be using it as it was designed. The communication tools may not be used effectively by everyone in the organization. These problems are, unfortunately, prevalent in organizations that have a high degree of autonomy in small subunits or divisions. The desire to allow operating entities the freedom to determine their own destiny sometimes backfires by allowing failure.

Some large corporations, therefore, have developed models for planning, communication, pay, and so forth, and find it necessary to survey employees on a regular basis to determine the effectiveness of these tools in totality and by entity. Corporations need to seek out "best practices" and be prepared to change and improve their models. Managers at local entities need to know how effectively the management tools are being used by different process groups.

Improve Leadership

The results from employee surveys can be used by corporate development organizations to improve the management tools and training used within their companies. The training programs can also help employees become more knowledgeable of existing tools. In either event, the results of these surveys can and should be used to contribute positively to the improvement of the leadership in the company.

REFERENCES

Byrne, John A. 1988. "Caught in the Middle." *Business Week* (3069) (September 12): 80–88.

Drucker, Peter F. 1988. "Leadership: More Doing Than Dash." *Wall Street Journal,* January 6:14.

Ford, Henry. 1922. *My Life and Work.* Garden City, NY: Garden City Publishing.

Kotter, John P. 1988. *The Leadership Factor.* New York: Free Press.

Labich, Kenneth. 1988. "The Seven Keys to Business Leadership." *Fortune* 118(9) (October 24): 58–63.

Perry, Nancy J. 1988. "Here Comes Richer, Riskier Pay Plans." *Fortune* 118(14) (December 19): 50–61.

Shores, A. Richard. 1988. *Survival of the Fittest: Total Quality Control and Management Evolution.* Milwaukee, WI: Quality Press.

Watson, Thomas J., Jr. 1963. *A Business and Its Beliefs.* New York: McGraw-Hill.

6

Customer Focus

Customer focus provides the business with continuous feedback from the competitive, dynamic environment in which it operates. This focus enables the business to provide products that continuously meet "customer needs."

CUSTOMER NEEDS

Customer focus is a process aimed at understanding customer needs and providing products and services that satisfy those needs. Being a leader in a dynamic, competitive environment requires a business to be better than its competitors in providing its products. "Better" means better products, better product price, and better product availability. "Better" does not always mean more product capability or lower price. It means optimizing the product attributes to be right for the customer's need and for a competitive situation; hence, the earlier definition; right product, right price, and right availability.

Focusing on the customer's needs as the primary basis for business strategy is the key to Japanese success. Focusing too much attention on competitive moves as the basis for strategy causes a business to adopt a reactive strategy. Competition introduces a new feature, your business reacts; they lower their price, you react; each reaction is too late, and your business is continuously one step behind. Focusing on customer needs is an offensive strategy. By painstakingly evaluating customer needs and developing business strategies around them, success can be achieved, and the head-to-head battle with the competition can be avoided (Ohmae, 1988).

G. Harry Stein (1986) identified innovation as one of the key elements that sustained the growth of the "survivors" over the years. Other studies have determined that successful innovators pay more attention to customer needs than their less successful competitors. A now classic study is the "SAPPHO" project conducted by the University of Sussex, in Brighton, Great Britain. Project SAPPHO (Scientific Activity Predictor from Patterns with Heuristic Origins) paired successful innovators with unsuccessful innovators, both seeking similar innovations. These pairings were established in two different industries (chemicals and instruments). SAPPHO collected data on many variables within these businesses to establish the statistical inferences about successful innovation. Of all the variables tested, understanding user (customer) needs scored highest. Second to user needs was marketing effort: market research, advertising, and user training (SAPPHO, 1980). These studies suggest that businesses must establish strategies, or processes, that ensure that adequate attention is paid to user needs when designing products and services.

The relationship between user needs and innovation is a fact that is demonstrated by the handful of companies that Stein refers to as "survivors." Hewlett-Packard, 3M, Merck, and Rubbermaid are just a few of the "elite" manufacturers who pay attention to customer needs in today's tough environment. They do so with painstaking attention to market research and customer contact. These businesses realize that the customers' productivity needs and their perceptions of the value of the product are the foremost important factors on which to focus.

Value

Receiving value is a fundamental user need. Products with the highest value offer greater benefit to the user at lower cost. For Yamaha Piano Company, adding value meant developing and selling digital adapters to make player pianos out of the 40 million pianos already sitting in homes (Ohmae, 1988). This strategy used technology to address a customer need and create a new market. Before Yamaha's innovative product, most owners of pianos were not getting the full benefit of them. Among the reasons given were that the kids may be grown and that the busy owners do not have the time to practice and play. Yamaha's piano technology allows prerecorded concertos to be played in the owner's home.

The Yamaha example illustrates the opportunity to add value to existing products with new products. Computer companies have pursued this market for years by offering new software, expanded memory, and faster processing speed, all retrofittable into old hardware as new technology becomes available. The unique thing about Yamaha is that it comes from an industry that is not used to thinking about such strategy. The people responsible for this strategy had to change their way of thinking and to think long and hard about their customers and their needs to design the right enhancements.

The thinking that goes into developing this type of breakthrough product strategy can all be for naught if the processes of the business are not efficient enough to sustain the advantage. Sustaining the competitive advantage requires a business to quickly introduce new products and enhancements, to have high-productivity manufacturing processes, and to have a continuous understanding of changing customer needs.

Today, many businesses are beginning to realize that real success in the marketplace requires breakthroughs in performance, quality, and price that are two or three times better than the competitive average. As Kenichi Ohmae pointed out, head-to-head battles with competitors only allow you to gain a temporary and marginal advantage. Major market-share gains require elements of vision, planning, and productivity in all functional areas to come together in major product breakthroughs.

These breakthroughs can only be achieved when an integrated approach is taken to understand customer needs and develop products. For example, marketing, design, and manufacturing must be equally involved in defining customer needs and "benchmarking" competitive performance. When all technologies are changing very quickly, marketing cannot afford to miss opportunities that possibly only the design team can see when visiting customers and studying competitors. The design team cannot keep manufacturing in the dark until the product has completed its design and then "throw it over the transom." An integrated approach means that these functions continuously work together at defining customer needs and designing products to meet those needs. This is the only way a business can break away from the pack and stay there.

PRODUCT QUALITY

A customer's buying decision is based on expectations created before the time of sale; these expectations come from many sources, including previous satisfaction with your product, experience with competitors' products, and advertising and sales claims about your product. After ordering a product, the customer will be satisfied to the extent that there are no disappointments. Therefore, the definition of customer satisfaction could be stated as the absence of disappointments or the absence of the elements of dissatisfaction.

In our dynamic, competitive environment, a perfect state of customer satisfaction never exists. Customer expectations continuously rise as technology and competitive offerings improve. Each business strives to improve as rapidly as the environment, but there is always some lag due to the time it takes to change product. The best a business can hope to do is improve faster than its competitors and thereby command a greater share of the market.

Defining product quality requires the business to have an understanding of user needs. Ultimately, a customer will be satisfied to the extent that those needs are satisfied; therefore, the broad measure of quality is the extent that those needs are satisfied. To define product quality, we look at customer needs and expectations, product attribute definitions, and specifications.

Customer Needs and Expectations

When buying a product, a customer has a set of expectations for that product. These expectations are neither satisfied nor dissatisfied at one particular time. The buyer's total satisfaction or dissatisfaction is determined over the life of the product. This process begins when the customer decides to buy the product. First, the customer must be satisfied that the product is going to satisfy the need. This decision is influenced by the level of communication that takes place between the seller and buyer. Next, the customer must be satisfied when the product is received and used. Last, the customer must be satisfied with the service if the product fails and with the alternatives when the supplier no longer provides the product. The period of time over which a customer owns a product and can be satisfied or dissatisfied is referred to as the customer life cycle. Let us consider some customer expectations over the customer life cycle.

Before the Sale

1. The specifications must be clear and unambiguous. They must relate to the application for which the product is intended.
2. It must be clear what regulatory requirements the product meets, that is, safety and health, etc.
3. It must be clear if the product will function correctly in the environment for which the customer intends it.
4. The delivery information must be reliable.
5. All product capability must be clear.

After Delivery

1. The product is received by the customer when promised.
2. The shipment contains everything expected and needed to use the product.
3. Operating and setup instructions are clear and complete.
4. The product functions as expected.
5. The product is received defect-free.
6. The product is easy to learn to use.

After Setting Up and Learning to Operate

1. The product continually meets specifications.
2. The product is reliable over time.
3. It is easy to verify continued conformance to specifications.

As the Product Ages

1. Preventative maintenance is clear, easy, and economical.
2. When the product fails, it is economical to repair.
3. Factory or service center repairs are handled promptly.
4. Spare parts are easily available at a reasonable cost.
5. Spare parts are available for the reasonable life of the product.

These are just a few of the expectations a customer may have of a supplier. Each business can enhance this list to reflect its customer base and should do so in an effort to understand any shortcomings. Once these expectations are known, a business must put in place the necessary organizational processes to satisfy them. When customer dissatisfaction does exist, an improvement process should be implemented to increase the level of customer satisfaction.

Quality Function Deployment

Understanding the needs of the customer and translating them into a set of design and manufacturing requirements are the purposes of quality function deployment (QFD). QFD originated in Japan and has only recently been used successfully by a few American companies. QFD is a process that has a dual purpose: first, it can be used as a planning tool to integrate the CEO's business plan with the strategies and responsibilities formulated at the lower levels; second, it can be used to translate customer requirements into internal product specifications (Sullivan, 1988).

The primary tool of QFD is the "house of quality," a correlation matrix used to define relationships between objectives and strategies. When used in the business planning process (hoshin or policy management, see Chapter 5), the objectives are stated on one axis and the strategies (means to achieve) are stated on the other. See Figure 6.1 (Sullivan, 1988). The roof of the house is the correlation matrix and is used to show the relationship between objectives, which in some cases points out conflicts in objectives. For an example, refer to Figure 6.1: the "X" in the roof shows the relationship between two objectives. By emphasizing the relative importance of achieving the objectives at different stages of the project, the design team can more effectively prioritize its efforts along the way.

Another use, and to date the most common application of QFD, is defining product requirements. When QFD is used properly, it helps companies design more competitive products, in less time, at lower cost, and at higher quality. By focusing on customer needs early in the design, the cross-functional team responsible for the development and introduction of the product finds that fewer engineering changes are required during development and after introduction. The product is also of higher quality in terms of being the "right product," which helps to optimize the price for the market. Japanese companies using QFD have shortened design cycle time by 30–50%, reduced engineering changes by 30–50%, and reduced start-up costs by 20–60% (Fortuna, 1988).

In this application of QFD, the house of quality is used to translate customer requirements into internal specifications. The customer

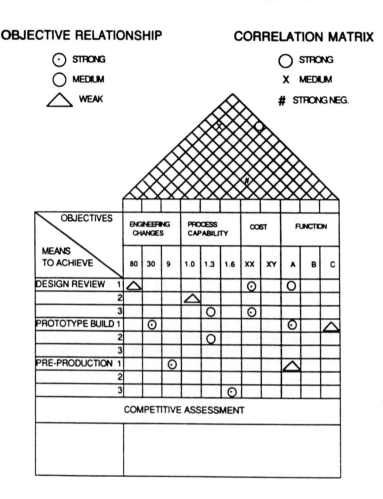

FIGURE 6.1 House of quality for policy management

requirements (objectives) are listed on the left and the internal specifications (strategies) are listed across the top. See Figure 6.2. The
resultant table/matrix is used to show customer requirements for features and quality and the relationship to the internal specifications.
For example, a commercial painter may require paint that "hides"
well (covers a bare surface with one coat) and spreads 300–400 square
feet per gallon. The paint manufacturer must translate the user's

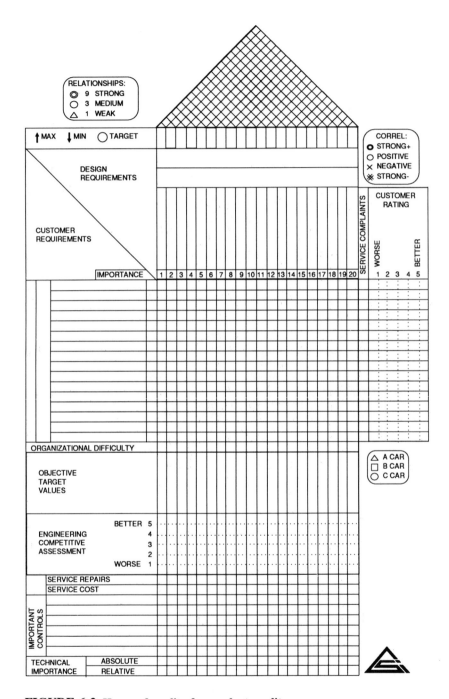

FIGURE 6.2 House of quality for product quality

SOURCE: Quality Function Deployment, three-day workshop, implementation manual, version 3.3, American Supplier Institute, Inc., Dearborn, Michigan, 1989. Reprinted with permission.

(painter's) requirements into product specifications for the type of paint: latex or oil-base paint, viscosity, and coloring agent content.

QFD is just beginning to show its potential. Companies like Ford Motor Company, who used it successfully with the 1988 Lincoln Continental, are continuing to evolve its application in their business (Sullivan, 1988). Other American businesses that are trying to push themselves to the head of the pack are also adopting QFD principles. QFD requires the same commitment to detail in the design process as statistical process control does in the manufacturing process. It is hard work, but the competitive nature of the environment is no longer tolerant of businesses that lack the discipline to apply these principles.

Product Specifications

Every product has a unique set of external and internal specifications. This often creates confusion about how to efficiently categorize product requirements. Much of the confusion can be eliminated, however, if businesses could standardize on a generic set of product attributes by which to define their products. A set of these definitions was proposed by me in my previous book. They were referred to as the FURPSAP measures, or product attributes (Shores, 1988). They are presented here for completeness.

The attributes on which customers make buying decisions, and which are the basis for evaluating the quality of a product, are called FURPSAP, which stands for the following:

Functionability

Usability

Reliability

Performance

Supportability/Serviceability (software/hardware)

Availability

Price

When defining the external or internal requirements of a product, each of these attributes must have at least one specification associated with it or the specifications are not complete. When using these categories with QFD, there may be so many specifications per category that it becomes necessary to make one product table for each attribute. This is not an unusual situation for a complex product.

The following definitions apply to an instruction and/or service manual for a complex product like a personal computer. The manual is referred to as the "documentation."

Functionability

The function of documentation is to communicate information about a product. The information that documentation normally needs to communicate is as follows:

1. specifications;
2. product features, capability, and intended use;
3. installation/assembly procedures;
4. operation/performance verification tests;
5. troubleshooting information;
6. operating procedures;
7. replacement parts lists;
8. repair procedures.

Usability

Usability means that the documentation is user friendly with easy access to product information: The documentation should be easy to read and follow. Typical considerations for documentation usability include:

1. table of contents;
2. index;
3. glossary;
4. prioritization/organization (start with first things first);
5. optimum quantity of information (not too much or too little);
6. correct literacy level relative to the intended users.

Reliability

Reliability is the accuracy of information: The accuracy of the documentation can affect the perceived reliability of the product if it misleads users on the operation, specifications, and diagnostic information of that product. The reliability of the documentation is determined by the following:

1. typographically correct information;
2. diagnostically correct (does it accurately identify failures?);
3. correct and validated operating procedures, installation procedures, etc.;
4. correct and unambiguous information throughout.

Performance

Performance is the contribution (time saved) to intended product use: Documentation that does not make a contribution to enhancing the product's use is wasted. As a matter of fact, it should be a challenge to the product designer to design products that are intuitively easy to use and save redundant documentation. Areas where documentation typically saves time are the following:

1. assembly and installation;
2. product learning cycle;
3. task accomplishment (productivity of use);
4. diagnostics/troubleshooting efforts.

Supportability

User costs to keep product and documentation current and operating include those related to repairs, calibration, preventative maintenance, and the costs to keep the documentation current. Elements of documentation that can influence these costs are as follows:

1. type of documentation update service provided; that is, change sheets, replacement pages, new manuals, etc.;
2. completeness of service instructions;
3. accuracy of parts lists and repair information.

Availability

Documentation availability relates to both new products and old. If a producer is in the habit of shipping new products without adequate documentation (preliminary documentation in some cases), the customer will not get the full benefit of the product. If customers lose old manuals, they need to have access to manuals that reflect the vintage of their product, not the new ones available today. The opposite is also true. When a product is updated, the documentation needs to be updated also.

Price

Documentation makes a definite contribution to the price of a product. It is not unusual in the software industry for an operating manual to represent more of the manufacturing cost than the product itself does. Documentation also makes a contribution to the customer's total cost of ownership, which should not be ignored. For example, if the producer is skimpy with service documentation, the customer could incur large repair costs over the life of the product. Consideration should be given to the various levels of documentation, the repair strategy, and the customer's total cost of ownership.

Each of the aspects of FURPSAP represents requirements that could and should have hard specifications defined early in the design process. When each product is defined in this way, the business is well on its way to realizing lower design costs, shorter introduction time, and fewer postintroduction changes.

PRODUCT DESIGN AND DELIVERY

The design team can now use the understanding it has of user needs and use QFD to set internal specifications. In the process of designing the product, the designers must focus on providing those specifications at the lowest possible cost of ownership to the customer, therefore, providing the highest value. The cost of ownership is defined as the cost of purchasing the product plus the cost of maintaining it. It could also include the productivity difference realized between this product and a competitive product, that is, two products may sell for the same price and have different support costs. If the unit with higher support costs includes better performance, making it more productive to use, it may still be the highest value.

The SAPPHO project also revealed that successful innovators were distinguished from failures by the efficiency of the research and development function. The study revealed, however, that efficiency was not necessarily determined by the least cost or the quickest introduction time. Innovation was defined not only as designing major contributions, but delivering them economically. One indicator of design efficiency was the number of defects introduced with the product release. Poor design quality and subsequent high manufacturing and service costs proved to undermine potentially successful innovations.

Product cost, quality, and delivery are directly influenced by the design of the product. The manufacturing process is affected by the number of parts, the quality of the design, and the match between the capability of the processes and the requirements of the product. Designing a product that conforms to the manufacturing process and is reliable at the lower cost is referred to as design for manufacturability.

Design for Manufacturability

In the pipeline analogy, the product design attributes can act as inhibitors to the fast and continuous flow of the product through the processes. Design for manufacturability (DFM) means designing a product so that there are no obstacles to throughput time. Although this definition for DFM has not been accepted universally, the logic of the definition will prevail as time-based manufacturing is better understood. Generally speaking, poor quality designs and higher complexity burden the manufacturing process with inventory and backlog queues, rework, test, and information transactions to raise the information-processing requirements. These factors cause longer throughput time and higher cost.

The design elements that affect queues and information requirements are in the following. They relate to the decisions engineers make every day as they go about their jobs. Engineers need to be willing to take personal responsibility for simplifying their designs to the lowest possible level and still meet the product specifications.

Design with fewest number of parts

Fewest number of different parts

Fewest number of processes

Fewest number of process steps

Fewest number of part suppliers

Product specifications do not exceed process capability

Designs that consistently meet specifications

Number of Parts

The total number of parts influences the number of assembly steps, the amount of MRP activity to keep track of the parts, and the in-process test time caused by higher complexity.

Number of Different Parts

This drives the number of inventory bins, purchase orders, purchasing people, MRP transactions, and many other manufacturing cost factors.

Number of Processes

This contributes to process engineering cost, equipment cost and maintenance, setup times, documentation, queues, etc.

Number of Process Steps

This drives setup times, amount of process information required, queues, and flowthrough time.

Number of Part Suppliers

This affects the number of purchase orders, purchase order tracking time, purchasing people, etc.

Designs Meet Process Capability

This results in less scrap, less product engineering time, less rework, and less queue.

Design Meets Specification Over Time (Reliability)

This results in fewer engineering changes, fewer production engineers, less warranty cost, less warranty information-tracking expense, fewer production holds, and lower queues.

When these DFM factors are given the attention they deserve throughout the design cycle, manufacturing costs will be as low as the technologies will allow.

I have never met an engineer who did not want to design a product to be as low cost and as high quality as possible. The problem is the amount of additional design cost to improve DFM. CAD and CIM tools can provide cost-effective opportunities to enhance DFM. One example is a printed-circuit design analysis tool developed at my division.

Our CIM efforts had already led to the development of a large data base for collecting and storing part and design information. This data base is a key part of our CIM efforts. The CAD system would transfer electrical design information to the HP Printed Circuit Design System (PCDS). PCDS is used by PC designers to design the physical layout of printed-circuit boards. When the design is completed, the physical design information, including material list, part numbers, and part locations, is transferred to a file in the data base (see Appendix D for details).

The PC design analysis tool has libraries of manufacturing information that specify whether each part number can be automatically inserted using existing processes. It also tells where in the process the part is inserted. If a part cannot be autoinserted for any reason, for example, wrong spacing, wrong hole size, or no tooling available, an exception report is printed. This information is passed back to the designer, who modifies the design, if possible. When the design is complete, loading pattern files are created that aid the autoinsertion programmer, and autotest programmers create, load, and test programs in less than a day.

The benefits to manufacturability are as follows:

1. Autoinserting is five times more productive than hand loading.
2. Autoinserting results in five times lower defects than hand loading.
3. The system increases autoinsertability of PC boards by 40–50%.
4. Prototype boards can be loaded using autoinsertion equipment and turned around to the design lab in one day.
5. First-run production boards can be assembled with automated loading on the day of design release.

There are other opportunities to gain similar advantages from CAD and CIM tools. Design tools, such as the HP ME 30 3D solid modeling product, provide many opportunities to improve manufacturability. The design data base contains all of the design information about interference fits, hole sizes, dimensions, and other design data. Libraries can be created that contain all of the tolerance capability of the manufacturing equipment. Comparison programs can then be run to see where a part or assembly will not meet specifications. This information can be provided to the designer to immediately allow for changes in the design.

As computers become faster, electronics design engineers will be able to simulate all possible component-tolerance variations. They will use libraries containing the tolerance information for each part number to run substitution tests on every possible combination of tolerances. This will enable engineers to introduce products with higher reliability, better performance, and fewer engineering changes. The possibilities for applying technology to improve design for manufacturability seem almost unlimited at this time.

Delivery Flexibility

Customers expect to receive their products in the earliest possible time. If not early, then on the date promised. There are two aspects to this product availability: one is new product introduction dates and the other is delivery of existing products.

New product introduction dates are the result of time to market and design cycle time. As discussed before, the time to market can be greatly influenced by the benefit of technology and information-processing tools. Computer integrated management has been very successful in electronically transferring design files to manufacturing processes so that manufacture of the product can start within the day of new product release.

The delivery of existing products can vary considerably, depending on order variability in product mix and volume. Businesses that are successful in maintaining short availability time without incurring huge and costly inventories are the businesses that are the most flexible. The "build-to-order system" discussed in Chapter 2 is a contributor to short delivery times and delivery flexibility. It allows businesses to keep delivery times as low as the production cycle time without finished goods inventory.

Flexible Manufacturing Environment

The idea behind flexible manufacturing environment (FME) concepts is to ensure that any mix of products within a reasonable and expected range can be processed through the factory with short cycle times, low setup times, and low inventory. To date, the most successful applications of FME have been in machine tool production shops. Machine tool shops are organized so that parts flowing through the shop are identified as needing some amount of capacity on each machine. The machines are numerically controlled in a fully automated environment. All machine capacity and capability are stored on a data base along with the part process information. As parts are pulled through the process, priorities are set to ensure that the capacity needed at the next process step is available. This process functions much the same as the JIT kanban process except the kanbans (cards) are not physically moved; they are seen in the computer.

In a less automated environment as exists in an electrical assembly manufacturing plant, the needs are at once the same and different. They are the same in principle, but different in practice. In this FME shop, products must share the same assembly and test hardware with access to unique software (documentation and automated assembly and test information). This requires setting up assembly and test areas around products that have common attributes.

In our manufacturing organization, we have up to ten different products being tested on the same assembly line using common hardware. These products move down the line one at a time and alternate by type and frequency throughout the day, week, or month, depending on customer demand. The assemblers have access to on-line computer documentation for each product and subassembly, which can be exploded to almost any level of detail. The test technicians access on-line test programs as required and run automated testing without intervention, unless there are problems. This type of environment is essential to short-cycle-time manufacturing using a build-to-order system if the product mix and the order volatility are high.

As these systems were developed, we found a need to also have a flexible work force, flexible in two senses of the word. The first is to be cross trained so as to move easily between work cells, depending on where the demand is on any given day. The second is to be flexible in terms of the number of hours worked in a given week. The former need for flexibility is satisfied by intensive cross training. The latter type required a different approach to employment.

The flexible environment and build-to-order system mean that the work load changes dramatically as the order level varies. In the HP environment, it is fundamental that we do not treat employment security lightly. We had to come up with a way of varying the work force without creating employment instability for the people to which we have committed full-time employment. We chose to set up a flexforce, which is made up of people who only want to work on a part-time basis part of the year. The flex-force assemblers now make up a necessary and substantial part of our total employee mix. We try to set the full-time equivalent, flex-force employee percentage to be equal to the normal order volatility percentage. This has been an extremely successful venture.

Ideally, the flex-force concept could also be extended to some of the project-oriented manufacturing overhead jobs. This would allow the potential for all people-related costs to be modulated with order and output levels. This would also enable businesses to buffer their permanent employees against recessionary cycles.

We have found that there are more than enough assembly people who want to work this type of schedule. They are usually second-income earners or retired employees. We have not, however, had much experience with this concept in the engineering, procurement, electronic data processing (EDP), and management functions.

Flexibility is becoming more in demand every day. It is being demanded by our customers and our employees. I have no doubt that all businesses will need to offer this kind of flexibility in the future. Flexible, short delivery of products is becoming an element of survival, not unlike quality and cost.

CUSTOMER SATISFACTION REVIEWS

Every organization needs people involved in collecting information about customers and customer satisfaction. Sometimes they are the people who sell the product, and other times they are part of a factory presales and postsales support group. Together they provide solutions to the problems that dissatisfy customers. The presales support group provides information to customers about products and collects information about customer needs. The postsales group serves to collect information about customer problems (dissatisfaction) and passes it back to the factory. The information is then used to identify changes that need to be made to the product or processes to improve the product and therefore improve customer satisfaction. See Figure 6.3.

Every employee has contributions to make, both to the improvement of the product and toward satisfying customer expectations. The following paragraphs are an overview of the contributions that different departments and people make to the improvement process.

Presales Support
The presales support function is part of the marketing interface responsibility. It encompasses customer contacts, advertising, and other forms of communication that set customer expectations. The market interface function (marketing) is also responsible for understanding customer needs, customer expectations, and competitive products. The information is collected, analyzed, and forwarded to the design group. The information is then used to define the external requirement spec-

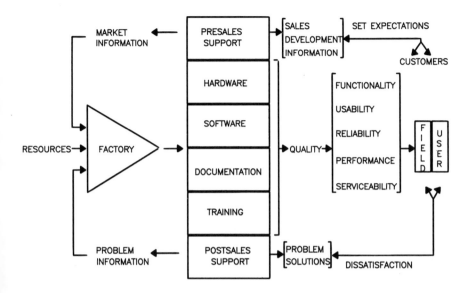

FIGURE 6.3 Customer satisfaction improvement model

ifications (ERS) of the next new products, thus contributing to a continued improvement of the product and, therefore, to customer satisfaction.

The Factory

The factory is made up of design, manufacturing, and support processes. All new product information is created in design and handed over to manufacturing when the design is complete according to the ERS. Manufacturing receives all of the documentation, specifications, and tooling required to consistently build and ship the product according to the ERS.

Manufacturing creates the processes to build the product as designed and to change the product as required by the changes in the environment. The changes may come from a number of sources, including changes in vendor part performance, part obsolescence, process changes, and, sometimes, product enhancements. The product shipped to the customer includes elements of hardware—the physical object—and possibly software (for computer products), along with some form of user information—instruction manuals, training, etc.

Customer Feedback Systems

Customer feedback systems may be formal or informal. The informal systems depend on the customers to call, write, or come to the business whenever they have a problem. The business may respond to solve the problem, but may or may not ever collect statistical information to help prevent the problem in the future. Most progressive businesses today are moving to more formal systems to collect customer feedback.

Formal customer feedback systems take many forms. They may be as sophisticated as a complex computer system or as simple as a customer reply card. In either event, the goals are the same: to collect as much information as possible about product failures, customer satisfaction, and opportunities for improvement. Once collected, the problems are analyzed for common causes and product improvements are planned.

Problem Analysis

The business has many people responsible for improving the product. The customer satisfaction data need to be collected, sorted, and distributed to ensure that the right information gets into the hands of the right people. Some of the information must go to the design lab, other information to marketing, and some to manufacturing. It is the responsibility of the people who operate the process, which is contributing to the problem, to use the analysis tools available to them to improve the product. These tools are discussed further in a later section.

Improve the Product

When the presales and postsales groups are both actively passing information along to the factory, and the factory is systematically changing the product, a state of continuous improvement exists for the product. If this product improvement process occurs faster than the customer expectations change, then a state of growth exists for customer satisfaction. It is important to point out that product improvement occurring at the same rate as the change of customer expectations does not result in the growth of customer satisfaction. A business that is continuously improving its product can still lose market share if the improvement is too slow.

It is also important to point out that this rate of change increases as the number of competitors increases. Technology is the driving force of this change and the rate of new technology development is proportional to the number of people contributing to that knowledge. For example, in an environment where there is only one business, change will be driven by the growth rate of investment that this business makes to develop new technology. The investment is limited by the revenue available, which is determined by the market size, or demand, for the product. Contrast this with the same environment but add two or three competitors. The total market size may not increase, but competitors will be compelled to invest higher percentages of revenue in technology development to gain a competitive position.

REFERENCES

Fortuna, Ronald M. 1988. "Beyond Quality: Taking Quality Upstream." *Quality Progress* XXI(6) (June):23–29.

Ohmae, Kenichi. 1988. "Getting Back to Strategy." *Harvard Business Review* 88(6) (November/December):149–156.

SAPPHO. 1980. *Success and a Failure of Industrial Innovation.* Brighton: University of Sussex, Centre for the Study of Industrial Innovation.

Shores, A. Richard. 1988. *Survival of the Fittest: Total Quality Control and Management Evolution.* Milwaukee, WI: Quality Press.

Stein, G. Harry. 1986. *The Corporate Survivors.* New York: AMACOM.

Sullivan, Lawrence P. 1988. "Policy Management Through Quality Function Deployment." *Quality Progress* XXI(6) (June):18–22.

7

Total Participation

Modern businesses are beginning to understand that people's ideas, attitudes, and actions are worth more when they are well focused on common goals through teamwork and collaboration. Individuals may be bright and make many contributions that differentiate them from the norm, but a group of individuals working together is always more effective than an individual working alone.

A simple example will illustrate this point. Write down a list of questions on any subject that would be difficult for an individual to answer correctly 100% of the time. Ask ten people to take the test and see what is the highest score obtained. Next, ask the same group to collaborate and reach consensus on each answer and then see what is the score of the group. In test cases of this nature, the group answer is usually 25% higher than the average of the individuals. Rarely will an individual score exceed the score of the group that collaborated. This simple example is indicative of the results that are achievable every day in work situations that require decision making. Teamwork and collaboration are the keys to achieving higher levels of success. The following discussions on total participation provide insight on how to achieve these synergistic results in business.

SYSTEM LINKAGES

The business is a large system of many people and processes. Working independently, the people will never become focused on doing the right things to achieve the greatest and best result. As in a tug-of-war, all people must be pulling together to win. A systematic means is needed to link the different people and processes together. In a tug-of-war, it is the rope that makes the link; in a business, it is the planning system used to create shared visions, plans, and decision making that makes the link. These are the tools needed to keep everyone focused and pulling together.

The Japanese *hoshin* system is used to achieve deployment of policy among all the employees. The process begins with the president of the company, who sets a very limited number of key goals for the year. These goals may be related as to quality, cost, and delivery. The people reporting to the president work as a team to understand these goals and develop strategies for achieving them. From each of them comes a new set of sublevel goals, and they work with their subordinates to gain consensus. This process is repeated across the entire company until each person has a hoshin stated for the year. In this way, everyone's goals and strategies are linked so that they are pulling in the same direction and at the same time.

As stated in Chapter 5, a business leader creates a vision by being perceptive of the changes in the environment and then visualizing (creating a metaphor) how the business can take advantage of those changes. Establishing the right linkages requires that everyone share this vision and be able to relate it to the job. An individual must look toward other teammates to help understand the common vision. The team (team is defined as an immediate work group that shares a common process and goals) must look to its manager for insight to understanding the relationship of the president's vision and the team's process.

Teams must meet regularly with their managers to review and renew their shared vision and plans. Managers need to continuously communicate the vision and seek to improve understanding and consensus. This is a continuous improvement process just as is every other process. The better everyone understands the vision, the better everyone's plans and actions will be in support of the goals of the business.

The Shared Vision

The vision is a statement about how the business expects to take advantage of the expected changes in the environment. Teams must plan to spend time together to review the current situation and understand what changes are taking place in and around their process. Some of these changes can be controlled locally; others cannot. It is important to know the difference, because those that cannot be controlled require a plan of reaction; those that can be controlled require a plan of prevention. Much of the information available to work with will come from outside the process, such as from customers; the rest must be created by the team members.

After the current and future situations are understood, the team should identify its purpose, expected contribution, and reaction to the expected changes. This may involve a considerable amount of brainstorming time (explained in Chapter 8) to identify all of the possible actions and then prioritize them. When this process is complete, the team will have a good understanding of the boundaries of change to be concerned about. They will probably relate to the following items:

Customer expectation changes

Products or service changes

Technology and process changes

Economic environment changes

Policy or philosophy changes

The most difficult part is creating the metaphor that describes what and how the team's process will need to function to be effective in the future as it has been defined. When the words are in place, the team has finished creating the shared vision. Some classic examples of visions include John F. Kennedy's 1961 speech of putting a manned spacecraft on the moon before the end of the decade, and Martin Luther King's speech of a future world of equality, "I Have a Dream."

Shared Plans

Shared plans are the specific actions the teams are going to take to make the vision a reality. Teamwork cannot fall by the side here because this is the time when the action begins. Teams must identify how they will make specific changes to quality, cost, and delivery. This involves changes to the product, the process, people skills (training), and policy. The amount of change in each area dictates the investment required and the allocation of resources.

Developing shared plans requires teams to get together and look at all of the possible alternatives. This includes analysis of the costs and benefits of the different paths available. (Analysis tools are discussed in Chapter 8.) The team must decide the following issues:

Which alternative actions will be selected?

What is the investment required?

What are the expected results?

How will the results be measured?

How is responsibility divided among the team?

When will the results be reviewed?

What are the rewards/risks?

How is success celebrated?

The answers to these questions form a statement of the team's shared plans. The following example relates to the printed-circuit assembly process. In this example, the team is made up of two people from day shift production and two from night shift. Both shifts work in the same process and they jointly developed a plan for improving the productivity of their process. This is a good example of team planning and the level of involvement and contribution assembly workers can have when working in a self-managed environment.

Add Kitting Project

(The Add area is where parts are added to printed-circuit boards after the wave solder process. This is a hand loading operation.)

Proposal:
Eliminate the kitting area, special request material (SRM), and the need for a person doing the kitting (kitter).

To eliminate the kitting area, we propose to put shelving above and below the existing Add track to hold the kludge kits.

To accomplish this, we are going to a one-bin system with trigger quantities (this is a JIT environment). We will also remove all

SRM from the kits and have storage cabinets at each work station with all SRM parts. This will make the kits smaller so they will fit onto the new shelves above and below the Add track, therefore utilizing less space in the Add area.

Having a one-bin system will eliminate the need for a kitter to refill the kits and put them away. The responsibility will then go to the individual adders.

Cost:

Cost of the new shelving that will go above and below the existing Add tracks. $746

Cost of SRM bins, which includes cabinets and clear boxes to store SRM: 8 bins $1,200

Approximate setup time and cost:
  Work Saturday overtime, 8 hours:
    One facilities support. $150
    Kitting team. $400
  Miscellaneous: labels for bins and resizing some kits
  (8 hours for kitting team) $264

Total cost $2,760

Savings:

Savings to Add area by saving kitting floor space (10 ft by 20 ft at $2.50/sq. ft/month) $500

Savings to Add area by not using a kitter (5 hr/day/month, approximately) $825

Monthly savings $1,325

Time:

We plan to have this project complete and operational by July 17.

Summary:

We are saving money by using less floor space and using less kitting labor (two-month payback with 600% annual return on investment).

We are reducing inventory in the Add area by going to a one-bin system.

We are making kits more accessible to Add people.

Consensus

Achieving consensus occurs when a group of people agree that they can jointly and mutually support an action. A team is not a team unless they have the ability to arrive at consensus on every issue. Unresolved issues create a waste of everyone's time and are counterproductive to the whole group. It is simply human nature not to support something with which you do not agree. By not resolving the issue, it will be debated endlessly, and actions will not be taken by some members of the team that are required for success. The Japanese use consensus very effectively. They believe that it takes more time to arrive at consensus, but when consensus is achieved, the results of the actions are many times more productive.

In a business, the existence of unresolved issues creates a backlog of doubt, uncertainty, and ambiguity. If the level of unresolved issues becomes too high, the entire system becomes ineffective, stress increases, and progress stops. The cost of what businesses refer to as "overhead" increases directly in proportion to the level of unresolved issues. The teams must work together to reach a mutual conclusion on each issue and support it as though it were their own choice.

Reaching consensus is not always easy, but it can be done in a very systematic manner. The first step is to clearly define the issue. Everyone in the group must be given an opportunity to contribute to the definition. This does not mean that the person who talks loudest prevails; it means that all persons are asked if they have something to contribute to the issue definition. Abstainers give up their right to complain or not support the definition later on. The next step is to brainstorm possible causes. Brainstorming means that everyone's ideas are accepted equally and written down without debate or evaluation. Ideas are listed, and efforts are made to ensure that every person has an opportunity to give unchallenged input.

The next step is to prioritize the opinions of the group as to the most probable cause. Prioritization is done by presenting valid data in a Pareto form to establish magnitude and importance. The next step is to again brainstorm possible action terms, and then again use a Pareto-type tool to organize the information. In the absence of good data, a more subjective method can be used. This method involves multivoting of the team members. Each team member casts a vote for the highest-priority item to make the selection. The previous steps have taken the team through several decision-making steps toward a final solution. When the group has completed this process, there is no longer any reason for an individual not to support the action. Those who deviate from this support without going back through the team's decision-making process should be eliminated from the team. This is a consensus process.

TEAMWORK

Teamwork has been discussed as it relates to creating a shared vision, plans, and decision making. This section discusses teamwork as it relates to quality teams, working with suppliers and customers, and employee suggestion systems.

Quality Teams

Quality teams are an integral part of business life in many successful Japanese companies. In Japan, quality teams, called quality control circles (QC circles), are a major part of corporatewide quality control and TQC. In response to increasing competitive pressure from these Japanese companies, many U.S. organizations began looking for the keys to Japan's success. What many saw were quality control circles. Some viewed this as a panacea for quality or morale problems. Many organizations began promoting quality circles. The results were mostly unimpressive. Quality control circles, under any name (quality teams, employee participation groups, etc.), did not seem to work very well. Many people rationalized that QC circles were a Japanese cultural phenomenon and would never be effective in the United States.

  This is not true. QC circles can only work in an organizational culture suited to participative management. This alone does not explain why QC circle programs often fail. The most important reason is simply that quality control circles do not function as a standalone program. They were never meant to. Davida Amsden (1983), a noted authority on quality control circles, wrote

> Circles are too often seen as a way to improve quality at the hourly level without middle or upper management's really believing in, let alone using, statistical quality control and other sound quality control methods. By contrast, quality control circles were the logical outgrowth of the Japanese realization that quality is everybody's job.

When used appropriately within the framework of quality control, quality control circles can be an integral and effective means of practicing quality control at every level of an organization.

Quality Teams Defined

A quality team is a small group of employees from the same work area who voluntarily meet to perform quality control activities. The group continuously, and as part of companywide quality control activities, solves work-related problems, implements or recommends solutions, and improves quality. A team consists of those employees who perform similar functions, generally for the same supervisor (all buyers, all assemblers, etc.), or it can consist of those individuals who, though they do not have similar responsibilities, work together on the same processes.

The reasons for quality team activities, carried out as a part of corporatewide quality control activities, are to:

contribute to the improvement and growth of the organization;

show respect for people (develop mutual respect);

build a productive and happy working environment (build teamwork);

exercise human capabilities fully (satisfy higher-order needs).

An organization involved in a quality control circle program should be "looking for an environment in which all workers accept the premise that quality is everyone's business" (Amsden, 1983).

Since quality teams are made up of people who are very familiar with a process, there is great opportunity for synergy when this pool of knowledge and expertise is focused on quality control and process improvement. The benefits derived from quality team activity include:

Improved quality, productivity, and communication.

Enhanced problem-solving skills and personal growth of the participating employees.

Increased job interest and ownership by employees who feel responsible and influential in their jobs.

Better teamwork and cooperation between individuals, groups, departments, etc.

Increased commitment to the team philosophy fomented by improved employee–management relations and communications.

More people involved in developing performance measures allow supervisors and managers to have access to more data than they could possibly develop without that assistance. This effectively increases the span of control management has by allowing it to more efficiently manage more processes.

Provides a basis for recognition for employees. Participation in a quality team provides members the opportunity to make significant contributions and to be recognized for them.

Self-Managed Teams

Self-managed teams are the latest in the team concept in participative management. They are the outgrowth of successful quality team implementation. When properly implemented, they provide benefits that go beyond quality circles. In the self-managed team environment, employees are encouraged to make many decisions within their groups that were previously reserved for management. These include, but are not limited to, the following:

Decisions on quality, cost, and schedules

Peer evaluation

Team hiring decisions

Process improvements

Setting team goals

Training peers

The motivational aspects of self-managed teams are similar to quality teams. Many people find the added responsibility to be rewarding and seek to take on more and more. Unfortunately, many employees do not do well in this environment. They feel threatened by having to do more, their stress levels increase, and they become unhappy in their jobs. They see the responsibility increasing as they take on supervisor's responsibility, but they do not necessarily see their pay going up proportionally. Self-managed teams may provide many opportunities for improving the contribution of the human resources of the business, but they also have many pitfalls that are not yet well understood.

Several American companies have had good results using the self-managed team concept. These companies include Ford Motor Company, NUMMI (a joint venture between General Motors and Toyota), Hewlett-Packard Company, and the A. O. Smith Automotive Products Company. These companies have implemented some form of employee–management participation system that encourages and rewards employees to work as a team and make more management-type decisions on behalf of quality and productivity improvements.

At the NUMMI plant, for example, they do not use quality circles, but they do have regular team meetings to discuss problems or opportunities to do the job better. Employees work with the Deming principles for quality control, using the simplest tools available for SPC. They use the "Andon" system, which allows employees to signal for help by pulling a cord when a problem on the line arises. If the problem is not resolved by the time the car moves to the next station, the assembly line automatically stops until the problem is fixed. Consensus is practiced by everyone at every level in reaching important decisions. Management's commitment to the workers has also increased, as evidenced by its decision to keep workers employed when the market was down. Managers and employees at the NUMMI plant believe in the system and take pride in what they are trying to do. The results are what could be expected: quality is higher, productivity is up, and employee grievances are down. All in all, this is a fairly successful experience with employee participation (Branst, 1988).

The A. O. Smith Automotive Products Company of Milwaukee was forced by competitive pressures to adopt a new management system beginning in the early 1980s. In its case, the unions were also a driving force to help bring about improvements in quality and waste reduction to protect jobs and preserve employee pay. A. O. Smith introduced the B.E.S.T. (Bringing Employee Skills Together) system. It is made up of a policy committee, an advisory committee, and teams. Participation from management and employees spans 1,566 employees and 90 teams at this writing. They have been seeking out and solving quality and productivity problems throughout the plant. Everyone is trained in problem-solving skills, and everyone shares in providing leadership.

The projects that they have undertaken include process, environment, safety, and product. The teams have reduced the changeover time for truck frames from over one hour to less than twenty minutes.

Other teams have made product and process improvements that have had savings in just about every aspect of the business. A. O. Smith has had a successful experience with employee involvement and plans to move forward with their B.E.S.T. system in the future (Ryan, 1988).

The Ford Motor Company's experience with self-managed teams was described to me by a Ford employee involved in the development of the team structure. Ford's Sharonville transmission plant experience dates back to a period in the early 1980s. The Sharonville transmission plant was aging, and the transmissions being made there were scheduled for obsolescence. The plant was scheduled to be closed, which threatened the loss of approximately 2,300 jobs. The local managers and employees, of course, did not want to see this happen, and so a management team spent several days in off-site meetings to come up with a new management system. Their goal was to improve the productivity of the aging plant enough to earn them the right to produce a new model transmission being introduced.

The new system they came up with embodied the principles of self-managed teams. At Ford these teams are called "business teams." Supervisors are trained to become advisors, and the teams are trained to solve their own quality and productivity problems. The Ford spokesman described the situation this way: Many employees were just coming to work, doing what they were told, and then going home without ever having to make a decision. Many of them would go home to family, community, and church, where they made responsible decisions that went far beyond what they did at work. The challenge was to figure out how to utilize all that human potential at work.

The Sharonville plant did improve quality and productivity, and it did earn the right to make a new transmission. The number of employees was reduced to half its previous level as a result of the many labor-saving and management-saving practices implemented. Teams met regularly to make decisions about quality, lay out new production flows, identify better tools, and make work-schedule decisions. The management organization became flatter, with fewer managers, and morale went up. Ford's experience with self-managed teams thus far is positive. It is proceeding to expand the concept into other parts of the corporation.

My experience with self-managed teams was at a division of Hewlett-Packard Company in Lake Stevens, Washington. We introduced the concept in December 1987 under the name of "Management by Objectives II" (MBOII). We had had a very successful experience with quality teams, and we wanted to take it a step further. MBOII was fashioned after the concepts of management by objectives, which strives for consensus of objectives, and then allows the teams to make all decisions at the lowest possible level.

The process was started by selecting an all-volunteer design team that represented employees from all departments and levels of management. In total, it was comprised of about sixteen people. The design team acted as a communication medium between the team and other employees in manufacturing and worked on operational issues: how to evaluate the success of the program, how to communicate progress, and so forth. The entire design team traveled to several other companies to see how they were implementing self-managed teams. The design team acted under the leadership of the manufacturing manager, who tried to ensure that the team also operated under the spirit of self-management; therefore, position power was discouraged and minimized.

The program was first introduced into a department that had about 125 employees who were responsible for assembling printed-circuit boards. The teams had already been trained in SPC through our quality team efforts, but they needed additional training in "working in groups" (WIG). WIG provided the tools for running efficient meetings, consensus decision making, and teamwork. As MBOII was implemented, we saw many of the expected benefits: defects decreased, rework went down, productivity went up, and the required number of supervisors went down.

As the implementation continued, we did begin to see some backlash from supervisors and employees. Much of this backlash came from people outside the pilot area. Supervisors became concerned when they felt that the primary objective was to eliminate their jobs. This was not directly communicated, but some of the changes that were first made were not always seen as being handled in the best interest of the employee; people's perceptions are always more important than reality. The second problem was with employees who simply did not want additional responsibility. We found that there really are people who are afraid or incapable of making the kind of decisions we were asking them to make. A third problem was pay. Hewlett-Packard Company has always had a merit-pay system, but some employees did not understand how the merit-pay system applied to the additional responsibility being "forced" upon them by MBOII.

The solution to these three problems, more than anything else, involved continuous and honest communication. The problems came about from a lack of mutual understanding, and communication was the only cure. We felt we needed to do more for the supervisors, so we put in place additional cross-training programs to qualify them for nonmanagement jobs in the same pay range. This required a considerable added investment, but a necessary one. If the supervisors do not support the program, there is not much hope that anyone else will. Further, the supervisors had to have confidence that we were not abandoning them. Continuous communication and assurances that they would not be left out in the cold were extremely important; these words, of course, must be followed by actions and integrity.

The employees who were set against additional responsibility and concerned about pay required a different tact. We had to approach these two problems together. First, we backed away from the strategy that all employees on the teams had to share equal responsibility. This gave the teams the discretion of deciding how much responsibility each team member would carry. It was necessary for some employees to change teams in order to find a compatible team. Every group of people is made up of individuals with varying degrees of ability and desire for responsibility. We had to recognize and acknowledge this in the team structure. Our merit-pay system played into this well, because it allowed employees to be ranked by their supervisor according to quality, quantity, teamwork, and willingness to take on additional responsibility. (We have not implemented peer evaluations.) Making these decisions and communicating them over and over until people understood them and believed them allowed us to move past these obstacles.

The benefits that we have seen are comparable to what others are reporting. In the year since implementation, we have flattened the organization by reducing one layer of management in the production area. The total number of managers was lowered by increasing spans of control. Our defect rates are down, and our productivity is up. We are moving forward with self-managed team concepts in conjunction with JIT, CIM, and TQC to build a more productive manufacturing environment.

Employee Suggestion Systems

An employee suggestion system provides another way to utilize the total potential of all members of an organization, in addition to quality teams and other participative management techniques (e.g., management by objectives, management by walking around). A suggestion system is the formal means by which individuals, groups, or quality teams contribute their ideas to the company. There has been a resurgence of the suggestion system in the last 10 to 15 years, but the form it has taken is more sophisticated and structured than before.

Benefits of Suggestion Systems

A properly administered suggestion program can provide many benefits to both the company and the employee. Member companies of the National Society of Suggestion Systems (NSSS) have reported savings of between $5 and $6.50 for every $1 invested in the suggestion program, with total dollar net savings of over $500 million among some 225 companies. In addition to the monetary advantages, there are other substantial benefits to be realized from a successful suggestion system:

Improved communication between workers and managers.

More team spirit in the company.

Employees more conscious of productivity.

Employees more conscious of process-improvement potential.

In addition to these organization benefits, an employee suggestion system can provide substantial advantages to the employees:

Opportunity to earn more.

Promotional visibility.

Peer respect and admiration.

Recognition for accomplishments.

Increased sense of contribution in the workplace.

Increased job satisfaction.

The objectives of a suggestion system fit in the philosophy of participative management. Having a suggestion system indicates that management considers everyone to be valuable. All must work as a team.

Elements of a Suggestion System

A large percentage of the suggestion systems implemented is either discontinued or is never successful. Unsuccessful programs invariably lack or mismanage one of the following vital features of successful systems. Administering a suggestion system requires a number of essential activities and features.

Top-Management Support. Complete support by management is necessary for the successful operation of a suggestion system for two reasons. Obviously, management must support the administration of the program. Also, management commitment affects the attitude of the rest of the organization toward the system. All management, including first-line supervisors, must be well versed in the program and must support it. Management must not criticize first-line supervision for not providing the suggestion themselves.

Supervisory Support. Supervisors are directly involved with the employees who are developing and submitting the suggestions. They must, therefore, be supportive of the program. Nothing kills suggestions more rapidly than supervisors who feel that making suggestions is an insult to their supervisory skills. They must understand the true nature of the program. Management must hold supervisors responsible for the operation of the suggestion system in their own departments. In addition to being supportive, supervisors can and should provide assistance to employees in preparing the suggestions.

Eligibility Guidelines. Clear policies should be developed concerning suggestor eligibility, scope, and type of suggestions, and how awards are determined and calculated. These policies must be structured to provide consistency throughout the company and over a long period of time, and must be published so that all employees are aware of them. All points of possible confusion or disagreement should be made as clear as possible before disagreements arise.

System Procedures. The processing regimen should be spelled out in detail. This should not only include routing conventions, but time guidelines, status requirements, and rejection procedures. It is extremely important that submitted suggestions be handled quickly. Lengthy delays in evaluating and communicating back to the suggestor dampens employee enthusiasm for the program significantly.

Suggestion Evaluation. There must be a consistent and equitable way of evaluating a suggestion. This must include the importance of the different considerations and some guidelines for comparison. A suggestion must be well defined so the employees know what constitutes an appropriate suggestion. The NSSS defines a suggestion as "an idea that poses a problem, potential problem, or opportunity; presents a solution; is written on the prescribed suggestion form; is signed by the suggestor; and has been received and stamped with the date and time by the suggestion office (or evaluator)."

Promoting and Publicity.   Aggressive promotion of the program is important for the ongoing success of the suggestion system to keep the employees up to date on what is being done and on what opportunities exist for them. Recognition and award ceremonies should be well advertised. Keeping the employees enthused about the program improves both the quality and the quantity of the suggestions made.

Record Keeping.   Detailed and accurate records should be kept of the specific suggestions made, the evaluation of those ideas, the implementation of suggested ideas, and the reasons for rejection of the suggestions that were not accepted. These data allow realistic reports to be made to management on the performance of the system. It should be understood that the submission of a suggestion is legally a contract between the company and the suggestor. The company can be held liable if the suggestion is rejected and the idea adopted without due credit to the suggestor. Accurate records of all improvement plans in the organization should be kept to safeguard against this happening.

Forms.   Suggestions should be submitted on a form designed for ease of completion and understanding. The forms should include enough information to make the suggestion clear and understandable.

Appeals Process.   Many suggestion plans have a process whereby a rejected idea can be appealed. This allows a suggestor to include more data or evidence supporting the suggestion. It also indicates the willingness of management to fully consider the ideas of the employees and to give the attitude that it is not infallible.

Rejection Process.   Some suggestions must be turned down. Rejecting someone's idea must be done very tactfully. The best way to do this is to make the contact personally and explain in detail why the idea is not feasible. Care should be taken not to couch the rejection in generalities or euphemisms and not to use the word "rejected." "Not accepted" sounds much more positive. The denial should be as positive and encouraging as possible. If the suggestion has already been submitted or is currently being investigated in the company, this should be made known to the suggestor with as much detail as is appropriate. This prevents the suggestor from feeling cheated if the suggestion is implemented at a later date.

Awards.   The most common award for accepted suggestions is cash. The amount of the award differs greatly between companies. The NSSS reports that on the average, the cash awards for cost-saving or productivity suggestions ran about 17% of the first-year savings obtainable from the suggestion. The range of cash awards was from virtually 0% to 100% of the first-year savings. Studies conducted by the NSSS indicate that there is a large difference in system success when the cash award is higher, with a significant change in perception when the award is above 10% of the first-year savings. Individual companies differ, and the larger organizations tend to have the larger awards. Regardless of the percentage, there should always be an upper limit on the award. The largest upper limit known by the NSSS is $250,000; most are in the $1,000 to $5,000 range.

Recognition.   In addition to the awards given for accepted ideas, it is important to give the employee recognition for the suggestion. This can be as large a motivator as a cash award. Many programs go so far as to have special dinners and publish the names of the winners in internal and external news sources.

Job Security.   The employees must be sure that no suggestion they make leads to someone losing a job because of an increase in productivity. The best safeguard against this type of deterrent is to have created a corporate climate wherein the employees do not fear job loss due to increased efficiency.

Supplier Participation and Involvement

Every company is a customer. A business benefits greatly when its suppliers provide products and services that consistently meet expectations. The higher the quality input to a manufacturing process, the higher the quality and productivity of the manufactured product. Managing the relationship with suppliers so that perfect parts are received is an integral part of total participation.

Benefits of Supplier Involvement

A supplier that uses TQC to improve its own manufacturing process is able to provide its customer(s) with better-quality parts, better service, and better price. Table 7.1 compares many typical benefits available to the business when suppliers use TQC to improve their quality (desired situation).

The actual, achievable level of quality probably lies somewhere between those two extremes. Still, the closer a business can get to the desired situation, the greater the improvements and savings in its own processes. The objective is to get rid of activities that exist to "inspect quality in" or to create "just-in-case" inventory.

The cost of low-quality materials from a vendor can be measured

TABLE 7.1 Present Situation vs. Desired Situation

Present Situation	Desired Situation
Parts seldom arrive on time.	Parts arrive on schedule.
Quantities received are often incorrect.	Parts arrive in the right quantities.
A portion of the parts are good.	All of the parts are good.
Many parts must be tested.	No need for incoming inspection prior to use.
Some bad parts reach production and must be screened by inspection and replaced by rework.	No component-related rework.
High quantity of safety stock to cover part failures, unreliable receipts, unacceptable shipments, and unknown rework quantities.	Little or no safety stock.
All items are counted in receiving or stores to verify ship quantities or they are assumed correct.	No need to count items to verify ship quantities.
Confirmations by purchasing staff in an effort to ensure delivery.	Little or no need for preexpedite contacts.

by summing the costs of the activities required to verify quality after receiving the materials—when it is too late for the vendor to do anything about the quality. These costs include but are not limited to the following:

incoming inspection;

rework to replace failed components;

servicing a higher warranty rate;

handling discrepant material;

carrying higher safety stock inventory in case bad parts are received;

lost sales due to customer dissatisfaction caused by product failures resulting from faulty components.

Late or early delivery from suppliers can also be costly. This is measured by summing the costs of everything done to compensate for parts nor arriving on time. Also added are the effects of not having the parts when they are needed. These costs include but are not limited to the following:

carrying higher safety stock inventory to prevent parts shortages before the next shipment arrives;

rescheduling direct labor and productivity lost due to the absence of needed parts;

lost sales due to customer dissatisfaction caused by delayed shipments resulting from back-ordered parts;

carrying increased work-in-process inventory—both to allow production rescheduling alternatives and to complete work with missing parts;

obtaining a replacement part through alternative sources (e.g., a distributor)—both the order-processing costs and the increased price.

A business cannot expect to minimize cost reduction through its own TQC efforts. The business depends on its vendors as well. No matter how much success is achieved internally, success is limited if it does not receive the right parts, on time, and defect-free.

Supplier's Customer Expectations
Quality is defined by customer expectations. The vendor must think of your business as its customer. The supplier's definition of quality must be identical to your expectations to provide you with quality parts and services. If these expectations have not been communicated explicitly, a vendor is likely to assume that they are similar to the following traditionally acceptable levels of performance:

Parts should be delivered before the due date (or at worst by a day or two later).

Delivery of parts as much as a few weeks early is acceptable.

A certain percentage of the parts will be defective. If the percentage is below a standard limit, we'll buy them.

The price will be as low or lower than is available elsewhere. The parts just need to meet a set of specifications on a drawing to be acceptable.

Accepting the traditional level of performance leads to inspection and rework. Even if a supplier is doing better and is giving the right part at the right time in the right quantity, holding to the traditional expectations just listed would still result in wasteful inspection and testing activities. Compare the traditional expectations to the following TQC-derived expectations:

Parts must be delivered by the due date and no more than two days early.

Parts must meet all requirements—zero defects.

Price should be lower (and if the supplier is using TQC, it probably will be), but price is not more important than quality and delivery.

The supplier must prevent problems before they affect our processes.

The supplier must provide us with documented proof, in the form of control charts on agreed-upon process quality measures (PQMs), that its process is in control.

The supplier is expected to assist us whenever its expertise can have a positive impact on our products or processes.

The parts should meet our implicit and unstated, as well as explicit, specifications.

This is a lot to expect. Therefore, it is absolutely necessary that companies work closely with their vendors.

Supplier/Customer Teamwork

There are many benefits to working with a supplier during the actual product design and manufacturing planning stages. Providing support in areas like quality control and technical assistance makes it easier for the supplier to understand and meet expectations. This level of involvement with a supplier requires a great deal of coordination between vendor and customer.

Involving suppliers allows them to use expertise developed through practice. Suppliers should be able to recognize potential improvement opportunities such as the following:

Overspecification for the use of the product;

Emphasis on original price versus cost of usage over the life of the product;

Emphasis on conformance to specification, not fitness for use; alternate designs or specifications that could improve performance or producibility.

Suppliers must be informed in detail about the process in which their product is involved and exactly how it is expected to perform. The most important specifications that a product must meet are performance specifications.

Not only must suppliers know the expectations of the product, but they must also be aware of delivery and process-performance documentation requirements. The object is for suppliers to build quality into the product by verifying the process. Process-quality measures applicable to the vendor process should be chosen, mutually agreed upon, and sampling methods specifically defined. As confidence is gained in the vendor's process, this documentation can formally take the place of incoming inspection.

It is important to maintain a teamwork attitude with the supplier even after the product is designed and being delivered satisfactorily. Any changes to the performance specifications or the expectations should be examined with the vendor as early as feasible. The supplier should be viewed, and used, as a vital resource in designing quality into products.

Communication and information sharing are vital elements to this type of supplier/customer relationship. This level of communication cannot come about overnight. The supplier and the customer must trust each other implicitly and explicitly. They must firmly believe that the other will never use any information in a deleterious way. This level of trust can only be built by experience. It takes a great deal of effort from both parties, but the potential benefits are well worth it.

It is important to realize that it is not necessarily desirable to develop a supplier/customer relationship to this degree with each supplier. Certainly, it is advantageous with suppliers of critical or high-dollar-volume parts. The level of interaction must be a function of the derivable benefits of success. This does not mean that suppliers of insignificant parts would not be given our expectations or that they would be excluded from any supplier training opportunities. The issue is always are the the available benefits worth the effort? Are the benefits commensurate with the dollars spent on the part, the production criticality of the part, the substitutability of the part, the uniqueness of the part, the technological level of complexity on the part, the current quality of the part, the level of interest expressed by the supplier, and the devotion to process improvement that can be developed in the supplier company?

Getting Suppliers Involved

TQC means change. Change is not always welcome. Suppliers may have difficulty accepting increased expectations at first. If they are not already involved in a TQC effort, they may initially see these expectations as unrealistic and impossible; and with their existing processes, this could very well be true.

A supplier may reluctantly agree to use TQC due to the pressure of retaining a company's business. This attitude must change for a successful adoption of TQC philosophy to take place. It must also be clear that this is not expected to take place immediately, but occurs over time with effort by both parties.

To help suppliers move toward and through TQC requires a huge effort on the part of the customer company. First, the customer must already be using TQC methods. Not only is example the best teacher, but extensive application of TQC methodology provides better definitions of the processes and more readily identifies those areas that vendors can influence. This allows more specific and accurate communication of expectations to suppliers.

The company should provide training in statistical quality control (SQC) and TQC to suppliers. Who should receive the training in the supplier organization depends on the situation. As a general rule, the higher the level of management that can be involved, the better. This rule should be tempered with the practicality of providing TQC training to management levels or supplier personnel drastically removed from the relevant supplier processes. The objective of the training is to introduce the supplier to TQC/SQC and to provide motivation for and guidance in using TQC. The vendor should be expected to pick up and carry the ball through its entire organization.

TQC training should start with formal lectures and seminars, but should evolve into a forum for practical application. It is a good idea for supplier training programs to have the same content as the internal training programs of the company business. This reduces duplication of effort in preparing materials and lends consistency to terminology and application between the supplier and the customer that aid in communication between the two.

Motivating suppliers can be very difficult, but is not impossible. Suppliers must realize that if progress toward meeting expectations is not made, the customer will look for alternative sources. Quality is required. It is imperative. A team effort is mandatory, so suppliers need to know that you will not abandon them if they are putting forth good effort. However, suppliers must also realize that if the effort is not forthcoming, other teammates will be found to replace them.

Suppliers should understand that TQC is a long-term commitment—on our part and theirs. Their participation leads to the benefits and advantages of TQC described elsewhere in this book and also ensures continuing and increasing their business as a supplier to the customer.

The first successes can lead to more involvement. This is true not only within one supplier's organization, but throughout the active vendor base. When a supplier applies TQC with successful results, that project should be advertised to the rest of the supplier's company. This rewards those who participated and motivates others. Success with vendors should also be shared with other suppliers (always with care not to reveal any confidential or sensitive information and only with their approval). This can be done with newsletters to all or selected suppliers, personal contact by the buyers, or by more formal means like supplier days (maybe even complete with award ceremonies).

Another motivation for suppliers to get involved with TQC is the opportunity to become certified.

Supplier Certification
Supplier certification is a means of certifying that a particular supplier is meeting or satisfactorily progressing toward meeting expectations. It means that the supplier is actively working on process improvement using TQC, can provide evidence of process control and improvement using control charts, and has reached minimum qualification levels for certification.

A supplier certification plan should "put the burden of producing quality components primarily on the supplier—where it belongs anyway." It should be demanding of the supplier. If certification is to mean something, it must be accomplishable, but not without effort.

Certification is done on a part/supplier or part-class/supplier relationship. It does not mean that a supplier is certified for all products, but, rather, indicates that a part or part-class is considered certified for that supplier.

The goals of a supplier certification program are as follows:

To improve the quality of incoming parts.

To provide positive feedback to suppliers.

To develop supplier performance to a level where incoming inspection is unnecessary.

To provide recognition to suppliers for accomplishments.

Suppliers and TQC

The philosophy of purchasing in a TQC environment is to create a situation in which the supplier is expected to perform in the same way as an internal process. Parts would be delivered directly to the production line just in time to be used and in just the right quantity. This situation is set forth as an ideal. It is not always achievable, but the closer it can be approached, the greater the advantages.

The goals of this type of purchasing, also known as "just-in-time (JIT) purchasing," are, among others, as follows:

To reduce lead times as much as is feasible.

To reduce order quantities as much as is feasible.

To reduce/eliminate purchased parts inventory.

To increase response to engineering changes.

To reduce paperwork.

To reduce/eliminate incoming inspection.

Supplier involvement TQC can lead to phenomenal process improvement. Suppliers should be treated as an extension of the customer organization and should be expected to perform as such. This requires an environment of teamwork and cooperation. The importance of this cannot be overstated. Process improvements lead to higher quality and lower costs.

REVIEW PARTICIPATION

Regular review of all processes and elements of the management system are necessary to ensure continuous improvement. For participation, the emphasis of the review should be on the success of the quality teams, supplier programs, employee suggestion systems, and the teamwork and participation of individuals. The following discussion focuses on the review of quality teams and individual participation.

The Quality Team Review

The result of the quality team's efforts is either an implemented improvement or a recommendation to management. Because the advantages of quality teams include improving employee problem-solving abilities and providing a basis for recognition, it is important that the quality team receive feedback on its efforts. The quality team review is an effective forum for both communicating necessary information to management and for providing a feedback opportunity for the quality team members. This feedback must not be confined to a study of the results.

Evaluating a quality team's activities should "emphasize factors such as the manner in which QC circle activities are conducted, the attitude and effort shown in problem solving, and the degree of cooperation existing in a team." Dr. Ishikawa (1985) gave this example of a weighted evaluation method:

Selection of the theme: 20 points

Cooperative effort: 20 points

Understanding of the existing condition and the method of analysis: 30 points

Results: 10 points

Standardization and prevention of recurrence: 10 points

Reflection (rethinking): 10 points

Total: 100 points

It is possible to structure quality team recommendations into an employee suggestion system. Even when this is done, the quality teams should be given the opportunity to formally present their activities and results. Having this review provides the following benefits:

Allows the quality team to present project results to management and customers.

Provides the quality team with the opportunity for recognition.

Gives management and customers an opportunity to see the project process, understand the recommendation, and ask questions.

Legitimizes management's concern and involvement in TQC and quality teams in the team member's eyes.

Allows management to own the TQC process through support of the quality team activity.

Provides feedback to the quality team on its activities so it knows the areas in which it can improve.

Gives good presentation and preparation skills practice to the members of the quality team.

Helps educate others by seeing what has been done.

Having a quality team (QT) review system consistent throughout a business allows for motivational contests, if it is structured enough to allow vastly differing projects to be compared. It also lends some consistency to the quality team efforts throughout the division. Continuous recognition for quality control efforts will help to perpetuate those efforts so that the program can remain active and effective.

Employee Reviews

Employees are frequently evaluated solely on the basis of the quality and quantity of their work. Unfortunately, teamwork is often left out and not considered an important element of the evaluation. The benefits of quality teams to the individual are many, and it is these benefits that should form the foundation of an individual's evaluation. When teamwork is working the way it should, the following individual characteristics prevail (Shores, 1988):

1. Employees share a feeling of mutual contribution.
2. Employees view management as supportive as opposed to commanding.

3. Employees demonstrate a higher level of responsibility.
4. Employees produce high-quality work out of desire, not compulsion.
5. Employees have more self-discipline toward team performance.
6. Employees feel more self-realization with the resulting job enthusiasm.

Quality circles provide the vehicle for developing this type of behavior. Quality circle reviews help teams understand their progress; employee reviews help individuals understand their performance. Annual evaluations and other elements of merit-pay systems should recognize these factors and be structured accordingly.

Improve Participation

Total participation involves many people and many processes. When the total system is linked together and working toward common goals, a business and its people have achieved the benefits of total participation. This requires the concerted effort of all employees and managers working together as teams and partners in excellence to achieve this objective. When reviews are conducted and the data tracked using the principles of SQC, the foundation for continued improvement of total participation is established.

REFERENCES

Amsden, Davida. 1983. *Quality Circle Papers: A Compilation*. Milwaukee, WI: American Society for Quality Control.

Branst, Lee, and Agnes Dubberly. 1988. "Labor/Management Participation: The NUMMI Experience." *Quality Progress* XXI(4) (April): 30–35.

Ishikawa, Kaoru. 1985. *What Is Total Quality Control?* Trans. by David J. Lu. Englewood Cliffs, NJ: Prentice-Hall.

Ryan, John. 1988. "Labor/Management Participation: The A. O. Smith Experience." *Quality Progress* XXI(4) (April): 36–46.

Shores, A. Richard. 1988. *Survival of the Fittest: Total Quality Control and Management Evolution*. Milwaukee, WI: Quality Press.

8

Systematic Analysis

Systematic analysis is used to ensure that the business is consistent in its response to the changes in its processes and environment. The presence of uncontrolled variation is the reason businesses buffer processes with inventories and the resultant long cycle times. Uncontrolled variation in quality is the reason for high levels of scrap and rework. This is the reason a systematic method of analyzing variation is required in every process. Every job—design engineering, assembly, accounting—must be studied for opportunities to reduce the uncontrolled variables in the process. This is the path to process improvement. The systematic approach to process analysis described here is a method that can be applied universally to identify those needed improvements.

First, recognize that every activity in an organization is a process. A process is a repeatable series of tasks that produce a product or service. For example, the assembly and test operations in manufacturing are obviously processes. But what about nonmanufacturing areas? In accounting, the financial books are closed every month using a definite process; the personnel department handles employee benefits according to a process; marketing engineers follow a process when responding to calls from the field; and R&D engineers follow a process for designing a new product. Every department in the organization depends on its processes to reach its goals.

Each department not only depends upon its own processes, but also on services from other departments. The next process (your customer) requires a quality product or service. People within the organization who depend upon your product are internal customers. The entire business is composed of producers and receivers, with each department playing both roles.

The entire business must be concerned with the ultimate customer, whereas every department within the business is focused on the needs of the next process. The process-analysis method presented here is customer-focused. The assumption is that the process exists to serve the customer and that nothing is more important than providing a better product.

The process-analysis method has many names, as discussed in this book. These include TQC cycle, Deming wheel, or PDCA cycle. The model used here is understand, select, analyze, plan, do, check, adopt, or the USA-PDCA cycle. The model begins with a focus on customer expectations and proceeds with an ordered set of process-analysis tools.

UNDERSTAND THE SITUATION AND THE ISSUES

To do so, ask the four customer questions:

1. Who are the customers of the process?

    Who are the people that depend on your product to do their job? A technician tests the product built by an assembler, but both depend on the quality of design from the R&D lab. Field engineers depend upon marketing engineers to answer their calls quickly and accurately. Marketing depends upon accurate financial statements and costings from finance. As previously stated, there are all levels of suppliers and customers within a business.

2. What products or services does the process supply?

    Is it an instrument, a part, a design, a report, information, forms, references, or what?

3. What are the customer needs?

    What needs is the product or service trying to meet?

4. To what extent does the product or service meet customer expectations for quality?

    How completely has the product or service met those customer needs? Are there any unmet expectations?

Select the Issue

Any difference between your product's performance and customer expectations constitutes an issue to be resolved. A single issue should now be selected for analysis from those identified in the "understand" stage. A simple statement clarifying exactly what is to be done is then developed from the issue. This is called an issue statement.

To aid in selecting the right issue to be addressed, compare all of the possibilities with regard to the following:

1. How much ownership or control a team has over the issue.
2. The urgency of the issue.
3. The trend: Is it getting worse or improving?

The issue statement specifies what is to be changed, the direction desired, and the process involved. The issue statement is the objective to be gained by doing the rest of the process analysis.

Analyze and Prioritize

This stage of the USA-PDCA cycle uses analytical tools to study the process and identify methods for improvement. These are the tools used in industrial engineering, statistical quality control, and other disciplines. Analysis is composed of two fundamental parts:

1. Current process analysis
2. Cause analysis

Begin the analysis of the current process by developing a flow chart of the way the process works now. Prepare the flow chart of the process at the macrolevel and develop additional detail as necessary. Sometimes a cause for a process problem becomes apparent when the process is detailed on a flow chart. If it does, skip on to the "plan" segment of the cycle and make the improvement.

It is often useful to do some "imagineering" by drawing a flow chart of the ideal process. Comparing the ideal to the actual flow chart can be eye-opening. Many people are surprised to find out just how many steps there are in their process. After the flow chart is constructed, identify quantitative measures of the process quality to assess the effects of action taken. A control chart is the ideal tool for tracking process-quality measures. If the measurements cannot be taken during the analysis cycle, a before-and-after study is recommended.

Finally, consider setting a goal for the quality measures. The ultimate goal may be perfection, but it is fun to set and reach intermediate targets.

CONTINUOUS PROCESS IMPROVEMENT

The search for causes of problems looks progressively deeper into the process. Tools used include the cause-and-effect diagrams, data-collection check sheets, Pareto analysis, scatter plots, and histograms. This step may include several iterations of cause-and-effect, data collection, Pareto analysis, and stratification to isolate the fundamental cause. Going through these cycles of analysis and improvement is what leads to continuous process improvement.

Data Collection

We have discussed the need to gather data for analysis as the process is improved. Next, the methods for data collection are discussed as a precurser to the methods used to analyze the data.

All data are the product of *a measurement system*. This system is composed of seven elements:

1. *The concept of the measurement being taken.* Initially, someone must identify the measurement and its interpretation. As an example, before you can measure the length of something, you must understand what it means for an object to have length. Although this may seem trivial for measurements of length, it becomes more challenging when talking about kinetic energy of an object or kurtosis (definition: the relative degree of curvature near the peak of a frequency curve, as compared with that of a normal curve of the same variance) of a distribution.
2. *Theory of measurement.* All measurement techniques are based upon a theory of why they work. This theory can be very basic or complex. A simple method of measuring the length of an object is by comparing it with a standard, such as a meter stick. On the other

hand, measuring the height of a tree cannot be done so easily. By using a stick of known length and walking out from the tree until, when held at arm's length, the stick appears to be of equal height as the tree, the height of the tree can be calculated using trigonometry.

3. *Equipment used.* The equipment used is a variable in the measurement system. Variations in the equipment cause variations in the resulting data.

4. *Environment under which the measurement is taken.* Imagine a forester measuring the heights of trees using the method just described. If the day is dark and raining, the measurements taken might well be different than if it were a bright sunny day with snow on the ground to give a clear sharp image of the tree our forester sights with a stick.

5. *Manner of recording data.* Consider two inspectors taking diameter measurements of a machined part. One takes the measure, then stops to record it with a pencil on a data-collection sheet. The other uses a special tool that electronically sends the measurement to the data base. Obviously, the second method would be preferable since the first presents opportunities for error in recording data.

6. *People taking the measurement.* Would two foresters measuring the same tree get the same height? It is unlikely. Any data are dependent on the person who took the data. Even two technicians using electronic tools for measuring the diameter of push rods must decide where along the length of the push rod the diameter is measured and how tightly the calipers are applied to the rod.

7. *Manner of taking the measurement.* The measurement technique, whether done carefully or haphazardly, affects the accuracy of the data gathered.

Change any one of these elements of a measurement system and the data change, too. It is important that everyone involved in a data-collection effort have a common understanding of the measurement system being used. Deming uses the term "operational definitions" (Deming, 1982). Such definitions of measurement assure the same meaning for everyone both today and next week. An operational definition implies many details and questions. It gives meaning to words like length, uniformity, and distortion, which have no meaning until expressed in terms of the measurement system.

Plan the Data-Collection Strategy

It is apparent that good data come from good planning. A good place to begin is by asking what one hopes to accomplish by collecting data. How, exactly, are the data going to help resolve the issue? Try to be as specific as possible. It is not too helpful to answer with, "if a bunch of data is taken, something should turn up."

Data-Collection Strategy

What data are needed? What quality measures? What related causal factors?

Do a thorough job of setting operational definitions.

Decide where in the process data should be collected. Use a flow chart of the process to guide you.

Having established the data-collection strategy, settle upon a sampling scheme. Sampling is usually better than 100% inspection due to limited resources. Statistical theory applied to sample data yields results reliable enough to make inspection of every piece unnecessary.

A decision must be made on how much data to gather. This depends upon the precision desired in the results and the expected random variation in the process. Initially, it is often good to take a reasonable amount of data to draw some initial conclusions before doing a technical sample-size calculation for later studies.

An issue related to sample size is the time period over which the data are gathered. This depends, to a large extent, upon the process cycle time.

Specifying how the data are recorded is the next order of business. Check sheets are an excellent tool for manual data-collection efforts.

Finally, decide exactly who is responsible for each area (data collection, data processing, reaction planning, etc.) before beginning to gather the data.

Statistical Quality Control

The term statistical quality control (SQC) causes anxiety for many people. The prospect of expending effort to take data and apply statistical formulas is not inspiring to most practical-minded employees.

Computers have assumed the burden of the mathematical calculations, so part of the trepidation associated with statistical methods should be eliminated. To some degree this is true, but often the complexity of calculations is not the reason SQC tools are left unused. The major obstacles are the collection of data and understanding how to apply the fundamental tools of SQC. This section illustrates the simplicity of applying these fundamental tools.

The goal of a process is to produce a product that is pleasing to the customer every time it is used. Today's product, tomorrow's product, and next year's product are to be of good quality. For the discussion here, the definition of quality is that the product meets the customer's expectations. Quality means fitness for use.

The customer expects perfection. This is nothing more than a rewording of the definition for quality just given. The term perfection may appear to be unrealistic, but it does make it clear that no deviations from expectations are to be tolerated. Customers do not want to hear excuses. A new car that leaks oil is defective, period. The customer does not care whether the problem is a design flaw or an assembly error.

The term perfection is used for a second reason; taken literally, it is impossible to attain. No human endeavor is capable of producing a product wherein every unit produced is identical to all the previous units and all units to follow. There is inevitably some degree of variation. This inevitable variation in quality casts serious doubt upon a producer's ability to provide the customer with perfection. Statistics is the study of variation and the first tool available for analyzing these problems.

When the meaning of "fitness for use" is dissected, another meaning comes out: "conformance to specifications." This meaning applies whether these specifications be expressed in inches, pounds, shininess, dB of distortion, smoothness, color intensity, clarity of picture, time to reply, or whatever. The customer's expectations eventually come down to a set of specifications that must be met. However, customers are not equally satisfied with a variety of products that meet the same specifications. Some products may perform better than the specification because of the individual quality characteristics.

SQC tools are used to understand the variation of quality, whether in specifications or out. The resultant trends may show that there is a set of problems causing the products to always be close to one edge of the specification. With this knowledge, it may be possible to change the process and improve the specification. For this reason, it is important to analyze trends that are favorable and unfavorable. Understanding why things are going so well may also be the key to future improvements.

Analyze Variation

There are two types of variation that may be present in a process: random and nonrandom. Random variation is always present, because no two items, processes, or circumstances are precisely alike. Nonrandom variation may or may not exist. A nonrandom deviation in product quality may be due to a vendor change, start of a new employee, uncalibrated instruments, or whatever. Nonrandom variation can be tracked down, the cause eliminated, and the process improved. This is not the case for random variation.

Consider a manager unaware of these two types of variation in product quality. The current month's scrap report is up from the previous month, due entirely to random causes, and our hypothetical manager demands an explanation from the responsible production supervisor. What could be said? Or, just as bad, this month's results are better than last month's and the manager seeks out the supervisor to offer praise. Again, what could be said? In a system under control, the supervisor could at least take solace in the knowledge that over a long period of time, praise will come as often as criticism when the manager does not understand SQC.

Assign Causes

The first objective of SQC is to identify the causes for nonrandom variation in product quality and eliminate them, thereby bringing the process under statistical control. Nonrandom variation may be beneficial or detrimental. The elimination of the cause may mean adopting a procedure that helps or eliminates something proven to degrade quality.

Achieving a stable distribution of product quality is not an end in itself. Doing so does not guarantee that the product meets specifications. The process of identifying and fixing the detrimental, assignable causes of the system variation to improve product quality and specifications is the goal of SQC.

The Tools of Statistical Quality Control

The following discussions on the tools of SQC are rather simplistic, in that they provide only a general explanation. There are many good books on the tools of SQC that delve more deeply into the methods of statistical analysis. This explanation is intended for the novice who only wants to know what the tools are, not necessarily how to use them in every application.

Control Charts

The principal tool for evaluating the state of control of a process is the statistical control chart. It is simple in construction and operation. It consists of an ordinary run chart enhanced by the addition of a center line drawn at the process average and lines drawn at the tails of the distribution so that if the process is under statistical control, 99.7% of all observations fall within the limits. See Figure 8.1. A certain history is required (usually, 15–30 samples) before a control chart can be constructed. This is necessary to get a reasonable estimate of the process average and variability about that average.

Four common problems in applying control charts are

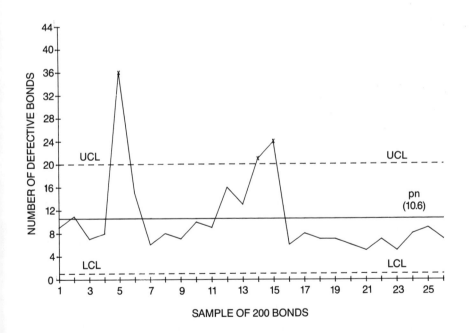

FIGURE 8.1 A control chart: The number of defective bonds per 200 attempts

1. selecting samples from the process;
2. knowing when to recalculate the control limits;
3. not using the chart real time (plot results often);
4. confusing specification and control limits.

Selecting appropriate samples for plotting is essential to the usefulness of a control chart. The samples should be tied closely to the working of the process. They should represent batches, days, shifts, machines, or whatever makes sense in the situation. Do not select samples in which there is a high likelihood of something disturbing the process within the time the sample is taken.

Recalculating the control limits with every sample is probably the most common error in the application of control charts. Once enough samples have been taken to calculate the average and control limits, extend them out into the future and continue to plot the data on the chart without adjusting the average or limits. In the special case where sample sizes are fluctuating, the control limits can vary with each sample, but the average should remain fixed. The process average and control limits should only be recalculated after an assignable cause of variation has been eliminated or a change made in the system itself, so as to make the control chart out of date.

Do not defeat the usefulness of this tool by failing to evaluate the most recent sample as it occurs (real time). It may be difficult to track down the reason for nonrandom behavior immediately. It is even more difficult if the trail is cold.

Computers have contributed to the last two problems with control charts. It is easy for software to recalculate the control limits with every sample and invalidate the chart. Also, electronic data bases are wonderful places to dump data for retrieval next week or next month, when it is too late to do anything. The very act of penciling in the day's point on the control chart causes a person to evaluate the chart for evidence of special cause and the need for action.

Finally, confusion over the difference between control limits and specification limits defeats the entire purpose of control-chart analysis. Specification limits are set by customer expectations for quality. Control limits are probably limits set by past performance of the process. The two are not related. It is for this reason that a process can be under statistical control but not meet specifications.

A common question regarding the application of control charts concerns the implications of a process out of statistical control during the time the first 15–30 samples are taken. If the process is under statistical control at that time, then the estimates of both the process average and variability are accurate. The resulting control chart accurately reflects the long-term process capability.

On the other hand, if the process is unstable at the time the initial samples are taken, the average and limits calculated are not an accurate reflection of the process capability. In fact, the average and limits may, when drawn on the data, look quite inappropriate. If this is the case, look for the cause(s) of the variation immediately. If a shift in the process occurred just prior to the taking of the data, the control chart may demonstrate statistical control but at a false average. In this case, the control chart does not detect the fact that the shift has occurred until the process shifts back again.

Histograms

A histogram is a bar graph of the process distribution. You may think of collapsing the control chart over the time axis to generate a histogram. See Figure 8.2. Note that if sample averages are plotted on the control chart, the distribution of individual observations has a wider spread than the distribution of sample averages. By drawing the specifications on the axis, the histogram clearly shows the position of the process relative to the desired performance. It is clear whether what is needed is a narrowing of the distribution, a repositioning of the distribution, or both. Remember, a histogram says nothing about statistical control; only a control chart does that.

Scatter Diagrams

The scatter diagram is, of course, the familiar X–Y plot, or correlation plot. See Figure 8.3. Whereas both the control chart and histogram plot a single quality measure, the scatter diagram requires a potential "cause" variable to be plotted against the quality measure.

Scatter diagrams can be enhanced by differentiating subsets of the data using various plotting symbols. Stratification of the scatter plot may turn up associations within subgroups that are quite different from the overall picture. This is only one way to make a simple scatter plot more useful. There are as many alterations to the basic scatter plot as there are people to plot them.

Pareto Charts

The common Pareto chart is a simple way of graphically displaying the relative contribution several factors make to the process under

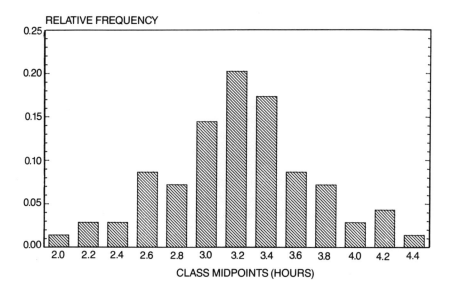

FIGURE 8.2 Relative-frequency histogram

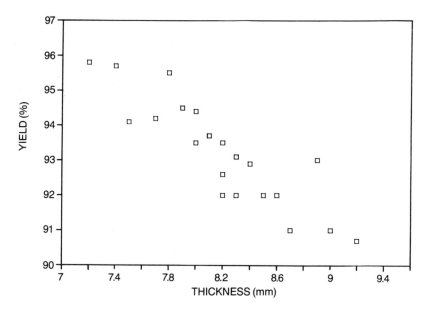

FIGURE 8.3 A scatter diagram: Wafer thickness vs. yield

study. It is a bar chart with the bars arranged from left to right in descending order; this tells at a glance which factor is the biggest contributor, second biggest, and so forth. See Figure 8.4. Care should be taken when choosing the criteria for ranking the categories, however. A Pareto chart of causes for rework might be based upon frequency of occurrence, dollars, or time lost. The biggest problem might well depend upon how one chooses to measure it.

The Pareto principle states that 20% of the problems drive 80% of the impact to the process quality. Pareto analysis then focuses attention on the vital few rather than the trivial many.

Cause-and-Effect Diagrams
Seldom in any analysis is there just one or two variables that are likely candidates for explaining the observed variation in the process quality. The cause-and-effect diagram is a good organizational tool for grouping and showing the relationships among the factors brought out in a brainstorming session. See Figure 8.5. The diagram is often helpful in designing a data-collection strategy.

Check Sheets
The check sheet is the principal tool for data collection. See Figure 8.6. A thoughtful layout of the factors taken from the cause-and-effect diagram can make data recording quick and simple. In fact, good planning can make data recording and data analysis virtually the same

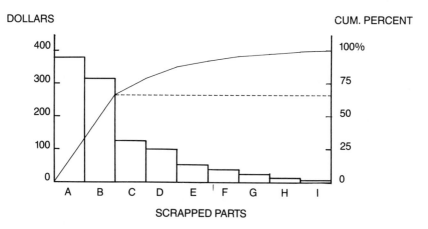

FIGURE 8.4 A pareto diagram showing cost (one month period)

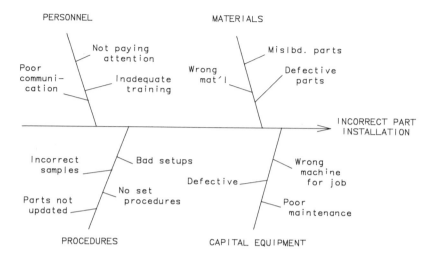

FIGURE 8.5 A cause-and-effect diagram for incorrectly installed parts

CHECK SHEET

PRODUCT: 03314–66503   DATE: April 1, 2001

PRODUCTION AREA: Board Test   DEPARTMENT: 8460

TYPE OF DEFECT: Open trace, lifted trace,   INSPECTOR: John Smith
warped board, internal short, other

REMARKS: New test bench installed today.   NO. SAMPLED: 500

DEFECT TYPE	CHECK	SUBTOTAL
OPEN TRACE	‡‡‡ ‡‡‡ ‡‡‡ //	17
LIFTED TRACE	‡‡‡ ‡‡‡ /	11
WARPED BOARD	‡‡‡ ‡‡‡ ‡‡‡ ‡‡‡ ‡‡‡ ///	28
INTERNAL SHORT	////	4
OTHERS	‡‡‡ ///	8
	GRAND TOTAL	68

FIGURE 8.6 A check sheet

step. For example, a histogram can be constructed from the hash marks made on a check sheet. By using a matrix layout and marking the check sheet with symbols rather than hash marks, multivariate data can be recorded in just one step.

Good planning in the layout of the check sheet can pay off in both the quality and quantity of the data collected. Collecting data can be a bothersome task and prone to error and neglect. Making the job easy reduces the chance of a data-recording error and lengthens the time people are willing to devote to the effort.

SQC Summary

It is easy to talk about data as quantitative information and lose sight that, in practice, data are the missing washer on a subassembly or the instrument that fails a distortion test. When the situation is in front of you, it is easy to believe that the most important action to be taken is to replace that washer or fix the instrument failing the distortion test. Of course, it is important to do these things. However, the most important action to be taken is to fix the process so that washers are not missing anymore and instruments no longer fail distortion tests. This is the goal that turns a missing washer into a data point.

The ultimate goal of SQC is total conformance to customer expectations. Sometimes businesses will conduct cost–benefit studies to justify the expense of making further quality improvements. These studies in themselves are additional costs of quality that go against the philosophy of TQC, that is, all productivity improvements must come through improvements in quality. Such cost analyses are often incomplete and overlook the new customers and increased market share that improved quality brings. As one business convinces itself that the improvement in quality is not worth the cost, another decides otherwise and captures the market. The quality potential must not be underrated. The "quality first" policy should be adopted without reservation.

The PDCA Strategy

Now that the situation is understood and issues have been selected and analyzed, the process-improvement cycle must begin. The assumption is that you have been able to assign a cause to the variation and now must take steps to eliminate the cause. In this way, the process is improved. PDCA is a universal method of improvement that can be applied to any process.

Plan the Action

Once a factor or set of factors is identified as being the cause of a process problem, brainstorm possible solutions. Choose a solution from your list and plan the implementation.

Do the Plan

Implement the solution. If possible, try the solution on an experimental basis rather than throughout the entire process.

Check the Result

Confirm that the solution had the impact intended. Check your process quality measures (PQMs) on the control chart or do a before-and-after study. If the results are favorable, go on to the *adopt* stage. If the solution did not work, loop back to the *analyze* or *plan* stage.

Adopt the Change

Once an improvement is demonstrated, incorporate it permanently in the process. Take care of all the details that sometimes handicap a good solution. Update documentation, conduct the necessary training, install the new tools, and explain to everyone the reasons for the change. Document the problem-solving process that led to the change, so that you can review it later should the need occur.

REVIEW THE PROCESS

This stage of the cycle is also a time for review. Assess where you are and where you have been. Has the process reached the goals set for it at the outset? Are there additional causes for process problems you have not addressed yet? You may want to move back to the *plan* or even *analyze* stage to develop and implement further process improvements. Iterate through the "plan, do, check, adopt" cycle as many times as necessary to achieve your quality goals.

Analyze the Process

After changes have been made to the process, it helps to continuously update the process flow chart. The team responsible for the process should meet regularly to ensure that they are each performing the process, as agreed. They should also brainstorm and analyze their effectiveness in utilizing the USA-PDCA process. Any problems in implementation should be addressed here.

Assigning Cause

Problems in implementation can come from many areas, including inexperienced people, management commitment, motivation, or any of the other elements of the management system previously discussed. The reviews need to be in concert with each other and causes of problems addressed in the proper context, for example, commitment should be dealt with as a commitment problem, not as a USA-PDCA process problem.

Improve the Process

The entire purpose of the USA-PDCA cycle is to identify the causes of variation in the process and eliminate them. By eliminating variation, the process is brought into control. This makes it possible to make accurate projections about the performance of the process. Accurate projections of the performance make it possible to accurately specify the product and meet the specifications without rework and scrap. When systematic analysis is used in a continuous cycle, the process gets consistently better.

REFERENCES

Deming, W. Edwards. 1982. *Quality, Productivity and Competitive Position.* Cambridge, MA: The MIT Press.

9

Improving the Management System

The task of management would be easier if it were simple. Unfortunately, the level of performance required of businesses in today's economic environment precludes an easy answer. In the past, the simple prescriptions to success included many of the singular parts of the management system described here. Management by objectives, TQC, leadership, and theories X, Y, and Z have all had their turn in the spotlight as *the* right solution to business management. Individually, these management tools are very powerful if done well and supported even by mediocre application of the other management tools. But that is not sufficient to be a leader in the current economic environment. Fifty years ago, success could be achieved by a fair amount of innovation and a poor management system. Twenty years ago, success required strong innovation and a mediocre management system. Today, success requires a high level of innovation and a high-performance management system.

Every business has elements of the total quality management system in place. They may be formal or informal, practiced by all employees or a few, done well or done poorly. In any event, a business needs to understand how well its management system is performing before attempting to improve it. Management is a process like everything else, and continuous improvement requires a thorough application of the USA-PDCA process as discussed in Chapter 8.

USA-PDCA

Understand the Situation

The process begins by "understanding the situation." The top managers in the business must have a comprehensive understanding of the current management system. This requires that a survey be done of the entire business. The survey should collect data and sort them in such a way that they show a map of the quality of the overall company management system, broken down by departments. The survey questions asked in the following cover the full range of the management system described in this book.

Identify the Issues

After the survey data are collected, they are then sorted onto a map, as shown in Figure 9.1. Each of the fifteen boxes has three scores resulting from the survey. There is an overall company map and a division/department map. The performance scores indicate where the management system is weak. The weakest areas are the issues with which to deal.

Prioritize the Issues

The three scores: performance, importance, and trend (PIT scores), are used to prioritize the issues. The highest priority is likely to be one with a combination of low performance, high importance, and low trend scores. The top managers involved in reviewing the scores can use some form of multivoting to set the priorities or they can use an algorithm, for example,

$$\text{Priority} = \frac{\text{importance}}{\text{performance}} \times \text{trend}$$

Ultimately, all issues must be addressed in the traditional fashion of continuous improvement. The first priority setting will establish the Pareto chart of items on which to work first. Subsequent surveys (possibly done annually) will rearrange the priorities.

Collect Information

The next step is to look in depth at the number one Pareto item. This will probably involve assigning someone the ownership of this issue. This person will need to thoroughly review all aspects of the company's efforts in this area and pull together a comprehensive report on the status. For example, the company's planning box under leadership may be the issue. An investigation is necessary to see if common planning tools are used throughout the company; to determine how the CEO's objectives are linked down and across the organization; to see if goals and measures are coherent across functions.

Analyze the Information

The information is pulled back together and reviewed by the management staff. Decisions need to be made about the kind of planning system the company wants to use. Ownership will have to be assigned to develop the "plan" for the new planning system.

Plan

The planning model needs to be developed; an implementation plan must accompany it, which will include training plans, a budget, resources, and a schedule.

Do

The plan needs to be implemented.

Check

Continuous progress reviews are necessary.

Act/Adopt

Problems are discovered in the reviews and acted upon. Changes are made along the way. The finished product is adopted and resources are moved to the next higher Pareto item.

ADMINISTERING THE SURVEY

The survey can be administered in two ways; the most comprehensive results are obtained by doing both. The first is a consultive survey. Here the consultant spends a day or more with the management staff and reviews pertinent information as asked for in the survey. This involves group meetings, one-on-one discussions, and a tour of the facility. The consultant scores all questions based on observations. A report is then generated that specifies the strengths and weaknesses of the organization.

The second method is to randomly select a sample population of employees (10–20%, depending on the size of the organization). The employees are each given a questionnaire based on the following consultive survey. The results are then tabulated and summaries prepared. This method gets directly to the perceptions and beliefs of the employees and provides top management with a view of what is really going on within.

When both methods are used, top management gets a top–down and bottom–up view of the organization's strengths and weaknesses. This is very powerful and necessary information. From here, top management can establish a plan of attack to strengthen its competitive fitness, organizational effectiveness, and customer satisfaction.

Consultive Survey

Scoring
For each survey question, the answers are numerically scored as follows:

Performance	*Importance*	*Trend*
5—always true	5—extremely important	5—rapid improvement
4—mostly true	4—very important	4—modest improvement
3—often true	3—moderately important	3—no change
2—seldom true	2—not very important	2—modest decline
1—never true	1—not important	1—rapid decline

MANAGEMENT COMMITMENT

Values

1. The organization has a written and stated set of values that emphasizes the importance and employee responsibility toward quality, customer satisfaction, teamwork, and business ethics.

PERFORMANCE	IMPORTANCE	TREND
☐ ☐ ☐ ☐ ☐	☐ ☐ ☐ ☐ ☐	☐ ☐ ☐ ☐ ☐
1 2 3 4 5	1 2 3 4 5	1 2 3 4 5

2. The values of the organization are well known, understood, and shared by all employees.

PERFORMANCE	IMPORTANCE	TREND
☐ ☐ ☐ ☐ ☐	☐ ☐ ☐ ☐ ☐	☐ ☐ ☐ ☐ ☐
1 2 3 4 5	1 2 3 4 5	1 2 3 4 5

3. The organizational and individual responsibilities for quality and customer satisfaction are clearly communicated and understood by all managers and employees.

PERFORMANCE	IMPORTANCE	TREND
☐ ☐ ☐ ☐ ☐	☐ ☐ ☐ ☐ ☐	☐ ☐ ☐ ☐ ☐
1 2 3 4 5	1 2 3 4 5	1 2 3 4 5

Investment

4. Training is a structured part of the environment. Time is dedicated to TQC activities as well as skill training. Formal training support is budgeted and provided in all areas. The business has a 5-year training and human resource plan.

PERFORMANCE	IMPORTANCE	TREND
☐ ☐ ☐ ☐ ☐	☐ ☐ ☐ ☐ ☐	☐ ☐ ☐ ☐ ☐
1 2 3 4 5	1 2 3 4 5	1 2 3 4 5

5. State-of-the-art computer-aided tools are utilized where applicable: computer-aided design, office automation, computer-aided testing, robotics, etc. The company has a 5-year capital-invest-

ment plan that focuses on utilizing the industry's current technologies.

PERFORMANCE	IMPORTANCE	TREND
☐ ☐ ☐ ☐ ☐	☐ ☐ ☐ ☐ ☐	☐ ☐ ☐ ☐ ☐
1 2 3 4 5	1 2 3 4 5	1 2 3 4 5

6. Automated inventory and information-management tools are utilized effectively throughout the organization. Material requirements planning, just-in-time production, and automated material-handling equipment are used throughout. The company has a 5-year inventory-management plan that drives toward zero inventory.

PERFORMANCE	IMPORTANCE	TREND
☐ ☐ ☐ ☐ ☐	☐ ☐ ☐ ☐ ☐	☐ ☐ ☐ ☐ ☐
1 2 3 4 5	1 2 3 4 5	1 2 3 4 5

Review Commitment

7. Top managers conduct reviews of the progress on the training, equipment, and material-management plans. These reviews are conducted at every level of the organization where investment decisions are made. Regular employee surveys are made to determine if the company's values are being adhered to throughout the company.

PERFORMANCE	IMPORTANCE	TREND
☐ ☐ ☐ ☐ ☐	☐ ☐ ☐ ☐ ☐	☐ ☐ ☐ ☐ ☐
1 2 3 4 5	1 2 3 4 5	1 2 3 4 5

8. The reviews are made on a regular basis, depending on the level in the organization. The president's audit might be annually, the next level semiannually, some quarterly, and some monthly, depending on the time frame of change expected. A plan exists for these reviews.

PERFORMANCE	IMPORTANCE	TREND
☐ ☐ ☐ ☐ ☐	☐ ☐ ☐ ☐ ☐	☐ ☐ ☐ ☐ ☐
1 2 3 4 5	1 2 3 4 5	1 2 3 4 5

9. The results of the audits and reviews are widely publicized and used as the basis for making improvements to the commitment of resources in the areas where they are needed most. All employees share in a feeling of confidence toward the continuous improvement of management's commitment. Employees reflect this in a commitment to the business.

PERFORMANCE	IMPORTANCE	TREND
□ □ □ □ □	□ □ □ □ □	□ □ □ □ □
1 2 3 4 5	1 2 3 4 5	1 2 3 4 5

LEADERSHIP

Planning

10. The organization has a formal written purpose and direction. It describes the market, the customer, and the products. It is broadly communicated and understood by all managers and employees.

PERFORMANCE	IMPORTANCE	TREND
□ □ □ □ □	□ □ □ □ □	□ □ □ □ □
1 2 3 4 5	1 2 3 4 5	1 2 3 4 5

11. The organization has a vision of its future. The vision articulates the future state that the organization is striving to achieve. It is motivating in a magnetic sort of way by attracting the enthusiasm of the people ·in the organization toward its achievement. The vision is broadly communicated and understood by all employees.

PERFORMANCE	IMPORTANCE	TREND
□ □ □ □ □	□ □ □ □ □	□ □ □ □ □
1 2 3 4 5	1 2 3 4 5	1 2 3 4 5

12. Plans are established at every level and part of the business. The plans are linked vertically and horizontally along the lines of the organizational structure. Quantifiable metrics are established with goals set that reflect customer needs. Strategies are developed with individual ownership assigned. Plans are jointly developed between teams and leaders and consensus is achieved.

PERFORMANCE	IMPORTANCE	TREND
☐ ☐ ☐ ☐ ☐	☐ ☐ ☐ ☐ ☐	☐ ☐ ☐ ☐ ☐
1 2 3 4 5	1 2 3 4 5	1 2 3 4 5

Motivation

13. Communication is open and continuous. Various forms of communication are used to enhance communication in three directions: up, down, and across. Regular update presentations are scheduled; coffee talks, management by walking around, and newsletters are used regularly. In general, there is an open air of regular communication and informed employees.

PERFORMANCE	IMPORTANCE	TREND
☐ ☐ ☐ ☐ ☐	☐ ☐ ☐ ☐ ☐	☐ ☐ ☐ ☐ ☐
1 2 3 4 5	1 2 3 4 5	1 2 3 4 5

14. The pay and other reward systems encourage quality and teamwork. Personnel evaluations and ranking systems emphasize quality, customer satisfaction, and teamwork. Opportunities for recognition are a structured part of the environment for achievement in quality, customer satisfaction, and teamwork. Employees view quality work as their goal, not an obstacle to higher output.

PERFORMANCE	IMPORTANCE	TREND
☐ ☐ ☐ ☐ ☐	☐ ☐ ☐ ☐ ☐	☐ ☐ ☐ ☐ ☐
1 2 3 4 5	1 2 3 4 5	1 2 3 4 5

15. Employees are empowered to act. They are allowed to make decisions consistent with their responsibility. Decisions are made at the lowest levels. All employees and managers feel free to act as required to improve customer satisfaction and quality.

PERFORMANCE	IMPORTANCE	TREND
☐ ☐ ☐ ☐ ☐	☐ ☐ ☐ ☐ ☐	☐ ☐ ☐ ☐ ☐
1 2 3 4 5	1 2 3 4 5	1 2 3 4 5

16. The leader reviews both the results and the process of improvement with the team on a regular basis. All members of the team

have a sense of knowing where they are and what future actions are important to their success. Priorities are continuously examined and adjusted.

PERFORMANCE	IMPORTANCE	TREND
□ □ □ □ □	□ □ □ □ □	□ □ □ □ □
1 2 3 4 5	1 2 3 4 5	1 2 3 4 5

17. Progress is measured and the data analyzed using a systematic process. Decisions are timely, and the progress of the group does not suffer due to indecision.

PERFORMANCE	IMPORTANCE	TREND
□ □ □ □ □	□ □ □ □ □	□ □ □ □ □
1 2 3 4 5	1 2 3 4 5	1 2 3 4 5

18. The results of the team reviews are used to continuously improve the quality of leadership. All employees feel a sense of ownership for the decisions made and the actions taken.

PERFORMANCE	IMPORTANCE	TREND
□ □ □ □ □	□ □ □ □ □	□ □ □ □ □
1 2 3 4 5	1 2 3 4 5	1 2 3 4 5

Customer Focus

19. The business has an understanding of customer needs. The business uses surveys, customer visits, and market reports to accurately segment the market, assess competitive strengths and weaknesses, and identify customer needs. The business has a continuous string of innovative products.

PERFORMANCE	IMPORTANCE	TREND
□ □ □ □ □	□ □ □ □ □	□ □ □ □ □
1 2 3 4 5	1 2 3 4 5	1 2 3 4 5

20. A structured process like quality function deployment (QFD) is used to translate user needs into product and process definitions and specifications.

PERFORMANCE	IMPORTANCE	TREND
□ □ □ □ □	□ □ □ □ □	□ □ □ □ □
1 2 3 4 5	1 2 3 4 5	1 2 3 4 5

21. Product goals are set to reflect specific measures of customer satisfaction based on the needs study. Product development and production of these products are measured against these goals.

PERFORMANCE	IMPORTANCE	TREND
□ □ □ □ □	□ □ □ □ □	□ □ □ □ □
1 2 3 4 5	1 2 3 4 5	1 2 3 4 5

Product and Process Cost and Quality

22. The business has an information-distribution strategy that focuses on simplification and throughput time for all processes. It has a 5-year information-management strategy that ties into the investment strategy.

PERFORMANCE	IMPORTANCE	TREND
□ □ □ □ □	□ □ □ □ □	□ □ □ □ □
1 2 3 4 5	1 2 3 4 5	1 2 3 4 5

23. The business is proficient in DFM. Product designs have clear and numerically defined goals for manufacturability. Products are released to production with few ongoing engineering change requirements. All engineers are committed to the goals of DFM.

PERFORMANCE	IMPORTANCE	TREND
□ □ □ □ □	□ □ □ □ □	□ □ □ □ □
1 2 3 4 5	1 2 3 4 5	1 2 3 4 5

24. Everything is viewed as a process. The manufacturing pipeline is designed for minimum throughput time. High-quality materials are demanded of suppliers, design documentation is accurate, employees are skilled and have all the information they need to do quality work. Defects are not tolerated, and continuous process improvement is pursued.

PERFORMANCE | IMPORTANCE | TREND

☐ ☐ ☐ ☐ ☐   ☐ ☐ ☐ ☐ ☐   ☐ ☐ ☐ ☐ ☐
1 2 3 4 5   1 2 3 4 5   1 2 3 4 5

Customer Satisfaction Reviews

25. The organization realizes that customer needs continuously change and, therefore, methods are used to track those changes. Customer surveys are conducted and regular feedback systems are in place to monitor customer satisfaction.

PERFORMANCE | IMPORTANCE | TREND

☐ ☐ ☐ ☐ ☐   ☐ ☐ ☐ ☐ ☐   ☐ ☐ ☐ ☐ ☐
1 2 3 4 5   1 2 3 4 5   1 2 3 4 5

26. The business has people and systems in place to analyze customer-feedback information and uses it in a continuous improvement strategy to improve its products.

PERFORMANCE | IMPORTANCE | TREND

☐ ☐ ☐ ☐ ☐   ☐ ☐ ☐ ☐ ☐   ☐ ☐ ☐ ☐ ☐
1 2 3 4 5   1 2 3 4 5   1 2 3 4 5

27. Customers expect responsiveness. The organization is responsive to fixing customer problems and responding to customer requests. The results of the presales and postsales analysis efforts are to create a continuously improving product as evidenced by improvements in delivery time, reliability, and other customer-satisfaction metrics.

PERFORMANCE | IMPORTANCE | TREND

☐ ☐ ☐ ☐ ☐   ☐ ☐ ☐ ☐ ☐   ☐ ☐ ☐ ☐ ☐
1 2 3 4 5   1 2 3 4 5   1 2 3 4 5

TOTAL PARTICIPATION

Organizational Linkages

28. Employees participate in planning. All employees understand the vision of the business and incorporate responsibility for achieving it in their plans.

PERFORMANCE	IMPORTANCE	TREND
☐ ☐ ☐ ☐ ☐	☐ ☐ ☐ ☐ ☐	☐ ☐ ☐ ☐ ☐
1 2 3 4 5	1 2 3 4 5	1 2 3 4 5

29. Every process is part of the whole and considered in the planning activity. Each area has a plan that is focused on its individual priorities and the contribution it makes to the whole.

PERFORMANCE	IMPORTANCE	TREND
☐ ☐ ☐ ☐ ☐	☐ ☐ ☐ ☐ ☐	☐ ☐ ☐ ☐ ☐
1 2 3 4 5	1 2 3 4 5	1 2 3 4 5

30. Measures, goals, and strategies are coherent. Vertical relationships are subsets of one another; horizontal relationships recognize shared responsibility. Consensus is achieved at every level in the planning hierarchy among supervisors and employees, and between team members.

PERFORMANCE	IMPORTANCE	TREND
☐ ☐ ☐ ☐ ☐	☐ ☐ ☐ ☐ ☐	☐ ☐ ☐ ☐ ☐
1 2 3 4 5	1 2 3 4 5	1 2 3 4 5

Teamwork

31. Quality teams or self-managed teams are a way of life. They are structured and have formal support and facilitation. They meet on a regular basis and receive the training and support to continuously improve their process. These groups are both permanent work groups that share a common process and ad hoc teams that meet to solve a common process problem but do not normally work in the same area.

PERFORMANCE	IMPORTANCE	TREND
□ □ □ □ □	□ □ □ □ □	□ □ □ □ □
1 2 3 4 5	1 2 3 4 5	1 2 3 4 5

32. An employee suggestion system is used to capture the ideas of others from outside their area of responsibility. The program is structured and provides rewards commensurate with the contribution.

PERFORMANCE	IMPORTANCE	TREND
□ □ □ □ □	□ □ □ □ □	□ □ □ □ □
1 2 3 4 5	1 2 3 4 5	1 2 3 4 5

33. Suppliers of materials and services participate in the TQC efforts. They are considered part of the team because their quality and delivery affect the productivity, quality, and customer satisfaction of your organization. Vendors are trained if need be and are certified based on their willingness and ability to work in this TQC environment.

PERFORMANCE	IMPORTANCE	TREND
□ □ □ □ □	□ □ □ □ □	□ □ □ □ □
1 2 3 4 5	1 2 3 4 5	1 2 3 4 5

Team Review

34. Forums are provided for quality team reviews. All employees are invited to attend, and top management is expected to attend. The selected teams review their accomplishments and processes used. This is viewed as an opportunity to learn for the attendees, and an opportunity for recognition of the team.

PERFORMANCE	IMPORTANCE	TREND
□ □ □ □ □	□ □ □ □ □	□ □ □ □ □
1 2 3 4 5	1 2 3 4 5	1 2 3 4 5

35. Consensus is continuously pursued. Managers and teams recognize that consensus is not necessarily compromise and that better decisions are reached when consensus is achieved. Decision-mak-

ing tools are used to help make consensus decision making an efficient process. These include brainstorming, multivoting, and structured analysis.

PERFORMANCE	IMPORTANCE	TREND
☐ ☐ ☐ ☐ ☐	☐ ☐ ☐ ☐ ☐	☐ ☐ ☐ ☐ ☐
1 2 3 4 5	1 2 3 4 5	1 2 3 4 5

36. Individuals are evaluated on their contribution to the team. It is recognized that the results of the team have priority over the results and recognition of any one team member. Personnel evaluations are structured to reflect this philosophy.

PERFORMANCE	IMPORTANCE	TREND
☐ ☐ ☐ ☐ ☐	☐ ☐ ☐ ☐ ☐	☐ ☐ ☐ ☐ ☐
1 2 3 4 5	1 2 3 4 5	1 2 3 4 5

SYSTEMATIC ANALYSIS

Understand the Situation

37. Situation analysis is a key part of all planning efforts. This analysis is comprehensive and the key to creating the vision that will drive the organization. This is a structured part of the planning process and articulated in the annual plan.

PERFORMANCE	IMPORTANCE	TREND
☐ ☐ ☐ ☐ ☐	☐ ☐ ☐ ☐ ☐	☐ ☐ ☐ ☐ ☐
1 2 3 4 5	1 2 3 4 5	1 2 3 4 5

38. A structured approach is used to identify the issues in the current situation that are not meeting expectations. Data are collected and analyzed through surveys if necessary.

PERFORMANCE	IMPORTANCE	TREND
☐ ☐ ☐ ☐ ☐	☐ ☐ ☐ ☐ ☐	☐ ☐ ☐ ☐ ☐
1 2 3 4 5	1 2 3 4 5	1 2 3 4 5

39. Clear priorities are set in the planning process. The focus is on the top Pareto items. All employees clearly understand what their priorities are.

PERFORMANCE	IMPORTANCE	TREND
□ □ □ □ □	□ □ □ □ □	□ □ □ □ □
1 2 3 4 5	1 2 3 4 5	1 2 3 4 5

Continuous Process Improvement

40. When issues are identified and prioritized, in-depth studies of the problem are conducted. Data are used to make objective decisions about what actions are to be taken as opposed to "gut feel."

PERFORMANCE	IMPORTANCE	TREND
□ □ □ □ □	□ □ □ □ □	□ □ □ □ □
1 2 3 4 5	1 2 3 4 5	1 2 3 4 5

41. The analysis tools are kept simple. SQC charts are an easy way for nonstatistical people to learn to analyze variation. Everyone uses these tools.

PERFORMANCE	IMPORTANCE	TREND
□ □ □ □ □	□ □ □ □ □	□ □ □ □ □
1 2 3 4 5	1 2 3 4 5	1 2 3 4 5

42. PDCA is used as the process-improvement model, and projects are documented in the PDCA fashion.

PERFORMANCE	IMPORTANCE	TREND
□ □ □ □ □	□ □ □ □ □	□ □ □ □ □
1 2 3 4 5	1 2 3 4 5	1 2 3 4 5

Process Review

43. Variation is analyzed in all processes. Data are continuously pursued and analyzed to establish the real reasons for variation. Few

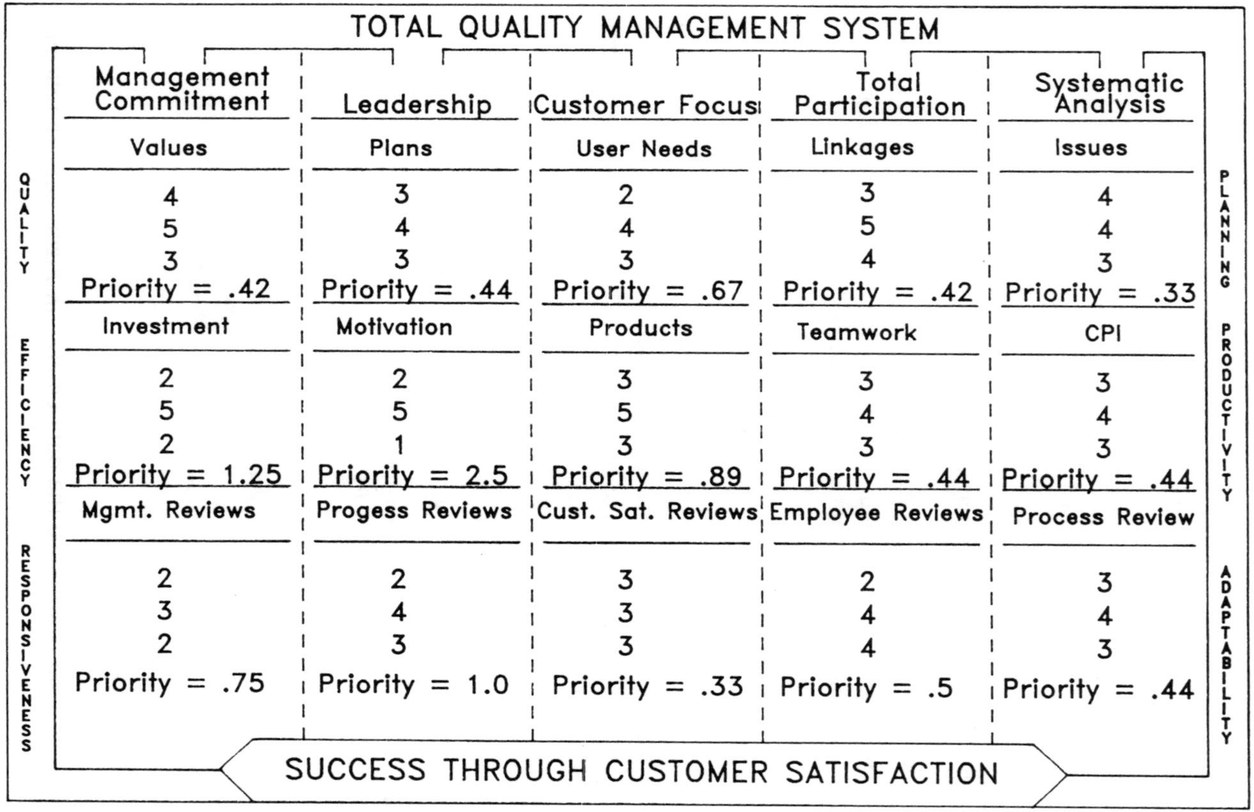

TOTAL QUALITY MANAGEMENT SYSTEM

	Management Commitment	Leadership	Customer Focus	Total Participation	Systematic Analysis
QUALITY / PLANNING	Values 4 5 3 Priority = .42	Plans 3 4 3 Priority = .44	User Needs 2 4 3 Priority = .67	Linkages 3 5 4 Priority = .42	Issues 4 4 3 Priority = .33
EFFICIENCY / PRODUCTIVITY	Investment 2 5 2 Priority = 1.25	Motivation 2 5 1 Priority = 2.5	Products 3 5 3 Priority = .89	Teamwork 3 4 3 Priority = .44	CPI 3 4 3 Priority = .44
RESPONSIVENESS / ADAPTABILITY	Mgmt. Reviews 2 3 2 Priority = .75	Progess Reviews 2 4 3 Priority = 1.0	Cust. Sat. Reviews 3 3 3 Priority = .33	Employee Reviews 2 4 4 Priority = .5	Process Review 3 4 3 Priority = .44

SUCCESS THROUGH CUSTOMER SATISFACTION

FIGURE 9.1 Management system perception map

decisions are made on the basis of gut feel or emotional responses to problems.

PERFORMANCE	IMPORTANCE	TREND
□ □ □ □ □	□ □ □ □ □	□ □ □ □ □
1 2 3 4 5	1 2 3 4 5	1 2 3 4 5

44. Analysis focuses on the good and the bad deviations from expectations; there is much to be learned from both. Causes for both good and bad variations are understood in all processes.

PERFORMANCE	IMPORTANCE	TREND
□ □ □ □ □	□ □ □ □ □	□ □ □ □ □
1 2 3 4 5	1 2 3 4 5	1 2 3 4 5

45. Analysis and improvement are continuously pursued in all processes and are a way of life for all employees and managers.

PERFORMANCE	IMPORTANCE	TREND
□ □ □ □ □	□ □ □ □ □	□ □ □ □ □
1 2 3 4 5	1 2 3 4 5	1 2 3 4 5

SUMMARIZE SCORES

The scores from this survey are to be sorted and summarized by category. There are three scores for each of the fifteen boxes. These scores are placed in the map in the order shown in Figure 9.1. The analysis then proceeds as discussed before.

Performance: 5 = always true

Importance: 5 = extremely important

Trend: 5 = improving rapidly

Interpreting the Survey

The results of prioritizing the data entered in Figure 9.1 are reflected in Figure 9.2. The priorities are based on calculations taken from a hypothetical business. The algorithm used is

$$\frac{\text{importance}}{\text{performance}} \times \text{trend}$$

In this example, employee motivation/morale is clearly the element most in need of improvement, followed by investment. The survey completes the first element (understand the situation) of the USA-PDCA cycle. The next step is to select the item to work on (motivation) and do some further analysis. Is the problem due to communications, review systems, or something else? After the analysis is done (which may involve group meetings with employees and a probable cause assigned), the PDCA cycle can be started to improve the result.

People frequently ask: How do managers get involved and practice TQC? The application of USA-PDCA to continuously improve the management system is clearly the most productive way.

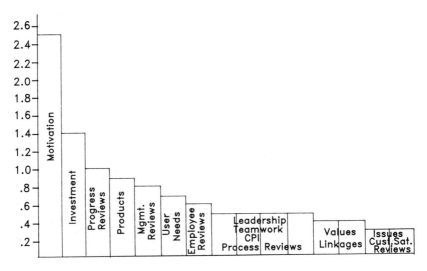

FIGURE 9.2 Importance/performance × trend

Part II

Appendixes

The following appendixes expand on the concepts introduced in the first nine chapters. The authors discuss TQC, JIT, automation, CIM, DFM, QFD, and human resource management. These papers were selected because they represent the best practices of some of the leading companies in U.S. industry.

A

"Six Sigma Quality" TQC, American Style

BILL SMITH
Senior Quality Assurance Manager; Motorola, Inc.

MOTOROLA, THE COMPANY

Motorola, Inc., is one of the world's leading manufacturers of electronic equipment, systems, and components produced for both the United States and international markets. Motorola products include two-way radios, pagers, and cellular radiotelephones, other forms of electronic communications systems, integrated circuits and discrete semiconductors, defense and aerospace electronics, automotive and industrial electronics, data communications, and information processing and handling equipment.

Ranked among the United States' 100 largest industrial companies, Motorola has about 102,000 employees worldwide.

As a leader in its high-technology markets, Motorola is one of the few end-equipment manufacturers that can draw on expertise in both semiconductor technology and government electronics.

FOUNDING AND EARLY HISTORY

The company was founded by Paul V. Galvin in 1928 as the Galvin Manufacturing Corporation in Chicago. Its first produce was a "battery eliminator," allowing customers to operate radios directly from household current instead of the batteries supplied with early models. In the 1930's, the company successfully commercialized car radios under the brand name "Motorola," a new word suggesting sound in motion. During this period, Motorola also established home radio and police radio departments; instituted pioneering personnel programs; and began national advertising. The name of the company was changed to Motorola, Incorporated, in 1947—a decade that also saw the company enter government work and open a research laboratory in Phoenix, Arizona, to explore solid-state electronics.

By the time of Paul Galvin's death in 1959, Motorola was a leader in military, space, and commercial communications, had built its first semiconductor production facility, and was a growing force in consumer electronics.

RECENT HISTORY

Under the leadership of Robert W. Galvin (Paul Galvin's son), Motorola expanded into international markets in the 1960's, and began shifting its focus away from consumer electronics. The color television receiver business was sold in the mid-1970's, allowing Motorola to concentrate its energies on high-technology markets in commercial, industrial, and government fields.

By the early 1980's, Motorola's communications and semiconductor operations each contributed about a third of the company's total revenues, indicative of fundamental strengths in electronic technologies at both the component and equipment levels.

An additional area of major corporate interest and activity, developed during recent years, has continued growth and investment in Europe and Asia, with particular focus devoted to Japan, where Motorola has established a wholly-owned manufacturing and marketing subsidiary, Nippon Motorola Limited, and a joint venture with Toshiba Corporation for manufacturing semiconductor products.

MANAGEMENT AND ORGANIZATION

The company's operations are highly decentralized, with business operations structured as sectors, groups, or divisions, depending on size. There are currently two sectors (Communications Sector, headquartered in Schaumburg, Illinois, and Semiconductor Products Sector, headquartered in Phoenix, Arizona), and four groups (Automotive and Industrial Electronics Group, headquartered in Northbrook, Illinois; General Systems Group, headquartered in Arlington Heights, Illinois; Government Electronics Group, headquartered in Scottsdale, Arizona; and Information Systems Group, headquartered in Canton, Massachusetts). In addition, there are groups within the two major sectors, and divisions within groups.

All of Motorola's businesses report to an Office of the Chief Executive, which includes: Robert W. Galvin, Chairman of the Board; George Fisher, Chief Executive Officer and President; and Gary L. Tooker, Chief Operating Officer and Senior Executive Vice President; John F. Mitchell and William J. Weisz serve as Vice Chairman and Officers of the Board.

DISTINCTIVE COMPETENCIES AND PRODUCTS

Each of Motorola's major operating units manifests a distinctive competence, which may be found in technology, distribution, management, or any combination of resources that allows the business to define why it is successful vis-à-vis competitors. All of Motorola's businesses are linked by a strategic commitment to high-technology electronic products.

The Communications Sector's distinctive competence is in meeting the needs of a diverse international market with highly specialized radio communications systems, many of which are built to customer order. These include two-way radio and radio paging products, and radio data transmission products.

The essence of the Semiconductor Products Sector's distinctive competence is an understanding and accumulation of skills in the design and manufacture of integrated circuits and discrete semiconductors that solve customer problems at low cost. Motorola builds a broad range of extremely complex, precise, and reliable microminiature devices.

The Information Systems Group provides networking capabilities for distributed information networks. This includes modems, multiplexers, and other data communications equipment required for managing information networks. The group includes operations of Codex Corporation (acquired in 1977) and Universal Data Systems.

The General Systems Group combines information processing expertise with knowledge of special technologies such as cellular radiotelephone systems. Products include mobile and portable cellular radiotelephones, cellular cell site and switching equipment and systems, and data processing equipment at the board, box, and systems levels.

The Government Electronics Group creates leading-edge technology, working in such frontiers as defense and space projects. Very complex, highly reliable systems result from this advanced ability. Motorola communications equipment has been used on most manned and every major unmanned U.S. space mission since Explorer 1 in 1958, including the sophisticated two-way communications gear used in Voyager's grand tour of Jupiter and Saturn, and the radio link that provided voice communication for man's first landing on the moon.

The Automotive and Industrial Electronics Group performs high-volume manufacturing of high-technology products, in automated environments, primarily for original equipment manufacturers. Its products include automotive powertrain, chassis and body electronics, sensors and power controls.

QUALITY AND PRODUCTIVITY

In an era of intense international competition, Motorola has maintained a position of leadership in the electronics industry through a combination of aggressive product innovation, strategic long-range planning, and a unique philosophy that allows each employee to contribute his or her insights to the achievement of quality standards. This philosophy is translated into action which brings work teams together to openly and effectively communicate ideas to help improve processes and products. PMP assumes that under the right conditions, employees will suggest better ways to do their jobs. Creating that atmosphere depends on a corporate culture inviting and requiring employee participation in managing the affairs of the company.

FACILITIES

Motorola has major facilities in ten states and Puerto Rico; and maintains more than thirty major facilities outside the U.S. Its employee population numbers more than 60,000 in the U.S., with the remainder found predominantly in Asia, Australia, Canada, Mexico, Europe, and the Middle East.

Major U.S. locations are located in and around Phoenix, Arizona, Austin, and Fort Worth, Texas, Fort Lauderdale and Boynton Beach, Florida, and Chicago, Illinois, where World Headquarters is located. In Japan, Motorola has facilities in Aizu Wakamatsu, Atsugi, Fukuoka, Kanagawa, Kumagaya, Nagoya, Osaka, Sendai, Tachikawa, and Tokyo, its Japanese Headquarters.

MOTOROLA, THE CULTURE . . .

To know a company's people is to know the company. The task of creating a "quality culture" is much easier within a company that already has an existing culture based on people values. A national quality award may be presented to a company, but it is the people within the company who, by their dedication to customer satisfaction, cause the company to be recognized as a quality leader. Loyalty to the company and to fellow workers, as well as respect for the individual, are essential ingredients to any quality winner, be it the Deming Prize or the Malcolm Baldrige National Quality award.

THE MOTOROLA FAMILY

Motorola is a public stock corporation, but the traditions established by Paul Galvin are carried on by his son, Bob, the current Chairman of the Board. As in many Japanese companies, there is a great spirit of loyalty to the company.

Many of Motorola's present employees are the sons and daughters of the original Motorola work force. A large number of families have more than one member working within the company. This feeling of the Motorola family is visibly evident at celebrations such as the anniversary or retirement of long service employees, or on the day of an employee's birthday, when his associates bring a cake from home to celebrate the occasion.

Bob Galvin recalled attending a retirement party held several years ago, where the retiree told him that he and his family had met the night before, and counted a total of fifty-two relatives who were either currently working for Motorola, or had worked for the company in the past.

Motorolans greet each other by first names, and unless visitors are present, the dress code is informal, that is, men do not wear suit jackets. It is not uncommon for an engineering department to declare a "jeans" day, especially if the day will be devoted to cleaning up the laboratory. After working hours, especially in the summer, many change into sports attire to play with their department's softball, volley ball, or other team on the fields of the "campus."

Visitors to the Schaumburg campus are occasionally surprised when they see an otherwise sane and rational engineer walking down the aisles with a large bowling ball chained to his ankle. This ritual began many years ago, when some engineers, as a practical joke, decided to literally attach a "ball and chain" to one of their colleagues who had just announced his wedding plans. Luckily, the originators of this custom were wise enough to attach a handle to the ball, and to make spare keys for the padlock.

Much has been written about the practice of life-time employment in Japanese companies. At Motorola, an employee becomes a member of the Motorola Service Club after ten years of service. A member of the Service Club may not be released from the company without the concurrence of Bob Galvin, the Chairman.

As changes in the business occur, the skills needed by employees may also change. The company is dedicated to the process of retraining long-term employees to accept new responsibilities in the organization. An Internal Opportunities System (IOS) exists to ensure that promotions are made from within the company wherever possible. All requirements for personnel must be posted on the IOS bulletin board, before applicants from outside the company can be considered.

PARTICIPATIVE MANAGEMENT PROGRAM

For many years, Motorola has sought the ideas of its employees. For about ten years, a process of participative management has been in place. All U.S. employees of the company who are not part of the Motorola Executive Incentive Program (MEIP) are members of a Participative Management Program (PMP) team. Teams are usually organized by function within an organization. Their purpose is to continually assess the process of performing their work, and to change it in ways which will reduce defects and reduce cycle time. The problem-solving efforts of these teams are directly analogous to Quality Circles. The quality and cycle time improvement rates for these teams are the same as the corporate goal, thereby providing incentives which directly support the quality improvement process.

BENEFITS

Motorola's Pension Plan now makes it possible for an employee who has worked for the company thirty-five years to retire at full pay, exclusive of any income he may receive from Profit Sharing equity. A day care reimbursement program is in effect for working parents, and an Employee Assistance Program is in place to provide professional counseling for whatever the problem may be, including financial, alcohol or drug dependency.

In addition, a contributory Profit Sharing plan has been in place since 1947, when the U.S. employees of the company voted to set up profit sharing instead of receiving a bonus. The Profit Sharing Council is made up of elected employees from all over the corporation.

THE QUALITY IMPROVEMENT PROCESS, 1981–1986
FIVE-YEAR TEN TIMES QUALITY IMPROVEMENT

In 1981, Motorola established, as one of its Top Ten Corporate Goals, the improvement of quality by ten times by 1986. The first reaction by some of the managers was that of skepticism. "We don't know how to achieve such an ambitious goal," they said. The response from the corporate management was, "We agree that it seems to be an impossible goal, but in the process of working towards this goal, we will find new ways to run our business at significantly improved quality levels. Each of these new ways will ultimately lead us to the ten times improvement."

And so, during the five years of the program, new methods were implemented. Some methods resulted in evolutionary improvement, but others resulted in step function improvements. The renewed emphasis on a "reach out" quality goal enabled Motorola to achieve its objective.

The most difficult problem which faced Motorola during this period was the fact that each organizational unit was free to define its own quality metrics. Within Motorola, a very decentralized company of many different businesses, it was a generally held benefit that each business was truly different, and so it made sense that each know the best way to measure quality for its business.

Because of the different way each business measured its quality level, it was nearly impossible for top management, in the normal course of conducting periodic operations reviews, to assess whether the improvement made by one division was equivalent to the improvement made by another. In fact, it was difficult for the manager of an operation to rate his quality level compared to that of another operation, because the measurements were in different terms. However, significant improvements were made regardless of the metric used. During the second half of 1985, with one and one-half years to go in this five-year program, the Communications Sector established a single metric for quality, Total Defects per Unit. This dramatically changed the ease with which management could measure and compare the quality improvement rates of all divisions. For the first time, it was easy for the general manager of one division to gauge his performance relative to the other divisions. They all spoke the same language.

THE QUALITY IMPROVEMENT PROCESS, 1987–1992
FOUR-YEAR ONE-HUNDRED TIMES QUALITY
IMPROVEMENT

Management of the Quality Improvement Process during this period is based on Motorola's practice of "Management by Measurement." This style of management says that by establishing measurements which are correlated to the desired end result, and regularly reviewing the actual measurements, the organization will focus on those actions necessary to achieve the required improvement.

In 1986 the Communications Sector adopted the uniform metric, Total Defects per Unit. In addition, because all operations were using the same measurement, the goal for defect reduction was uniformly applied to all operations. The required percent reduction was the same, regardless of the absolute level. The improvement rate achieved by the Sector was much greater than had been achieved in the five-year ten times program, and so the measurement was adopted by the entire corporation.

In January 1987, Motorola restated its corporate quality goal to be:

Improve 10 times by 1989

Improve 100 times by 1991

Achieve Six Sigma Capability by 1992

This goal applied to all areas of the business, not just product quality.

COMMON QUALITY METRIC

The use of the common metric, Defects per Unit, at last provided a common denominator in all quality discussions. It provided a common terminology and methodology in driving the quality improvement process. The definition was the same throughout the company. A defect was anything which caused customer dissatisfaction, whether specified or not. A unit was any unit of work. A unit was an equipment, a circuit board assembly, a page of a technical manual, a line of software code, an hour of labor, a wire transfer of funds, or whatever output your organization produced.

And thus, Motorola's Six Sigma Quality process had reach maturity.

THE MALCOLM BALDRIGE NATIONAL QUALITY AWARD

Although it does not have the many years of tradition associated with the Deming Prize, the ultimate goals of the Malcolm Baldrige National Quality Award are the same as for the Deming Prize.

Having visited several Deming Prize winners, as well as having been through the Malcolm Baldrige award process with the first winner, I will note some differences between them.

The Deming Prize process appears to be much more formal, including several years of "apprenticeship" in preparing for final consideration as a winner. Many years may elapse between the original application for consideration and the awarding of the Deming Prize. The Baldrige Award has no such waiting period. In both cases, the company must establish and maintain an effective quality improvement process, and in either case this is a process which takes years to accomplish.

When one visits different Deming Prize winners, and views the flow charts and plans which govern the quality improvement process, one perceives a great deal of similarity, even to the format of the documentation of the processes. This is perhaps due to the use of consultants from the Japanese Union of Scientists and Engineers (JUSE), whose standards are uniformly applied to all applicants. This degree of standardization is not yet apparent in the Baldrige process, although it may evolve as the award process matures. Currently, there is no standard requirement for format of the documented process, although the application criteria are fairly specific on intent, execution, and results.

As time progresses, some changes may occur in both processes. Each will recognize some advantage to using a technique employed by the other. Since the quality improvement process itself is never ending, we can assume these changes will be adopted as customer expectations increase.

The Award

The Malcolm Baldrige National Quality Improvement Act of 1987 was signed to law on August 20, 1987, by President Ronald Reagan. This law establishes an annual United States National Quality Award. The purposes of this award are to promote quality awareness, to recognize quality achievements of U.S. companies, and to publicize successful quality strategies.

The award program is managed by the National Institute of Standards and Technology (formerly the National Bureau of Standards), United States Department of Commerce. It is administered by the Malcolm Baldrige Quality Award Consortium, a joint effort of the American Society for Quality Control and the American Productivity and Quality Center. All funding for the program is from private contributions.

The award, a gold medal embedded in a crystal sculpture, formally recognizes companies that attain preeminent quality leadership. It is intended to encourage all companies to improve their quality management practices in order to more effectively compete for future awards. The winners of the award are obligated to widely disseminate non-proprietary information about their quality strategies to other American companies.

As many as six awards may be given each year. Up to two awards may be given in each of three categories: (1) manufacturing companies or subsidiaries; (2) service companies or subsidiaries; and (3) small businesses of 500 employees or less. Fewer than two awards will be given in each category if the high standards of the award are not met. This was the case in 1988, when no award was presented in the service category, and only one was presented in the small business category.

Should an entire company win, that subsidiary may not apply for five years, but another subsidiary, or the entire company may apply the following year.

1988 Winners

The winners for 1988 were Motorola, Incorporated, as an entire company, the Commercial Nuclear Fuel Division of Westinghouse Electric Corporation, and Globe Metallurgical, Incorporated, a small business specializing in ferroalloys. All are manufacturers. No service companies met all of the exacting standards of the award criteria.

Application Process

Each company must submit a written application, not to exceed seventy-five single pages, covering seven major subject areas. In the case of a large diversified company, such as Motorola, a supplement, not to exceed fifty pages, may be used to provide information on each of the essentially different businesses within the company.

Motorola did, in fact, submit a basic application which answered all seven subject areas on a company-wide basis. Four supplements were also submitted for the Communications, Cellular Telephone, Semiconductor, and Automotive businesses.

Examination Process

All applications are numerically scored by a Board of Examiners. The Board of Examiners is comprised of more than one-hundred quality experts selected from industry, professional and trade organizations, and universities. Those selected must meet the highest standards of qualification and peer recognition. Examiners take part in a preparation course based upon the examination items, the scoring criteria, and the examination process.

Each application is scored by at least four members of the Board of Examiners. A maximum score is 1,000.

High-scoring applicants are selected for site visits to be made by one or more teams of examiners. The primary objective of the site visits is to verify the information provided in the written application and to clarify issues and questions raised during review of the application. The site visit may result in lowering the original score, leaving it unchanged, or, in rare instances, increasing the score.

A complete written evaluation report, along with the examiners' recommendation is then forwarded to the Board of Judges.

Award Process

A panel of nine judges from the Board of Examiners reviews all data and information and recommends award recipients. The judges' award recommendations are based not only upon scores applicants receive on the written application, but also on the judges' assessment of overall areas of strength and areas for improvement as determined from site visits. All applicants receive a written feedback summary relative to the award examination categories.

THE QUALITY IMPROVEMENT PROCESS

Highlights of the Motorola quality improvement process are outlined in the following paragraphs, as they relate to the seven major subject areas. Naturally, all of the tools for quality improvement have been common knowledge for many years. The key to a successful quality improvement process is not in the tools themselves, but rather in the pervasive use of those tools within the everyday conduct of business.

The success of Motorola can be attributed to a clear understanding by management, from the Chairman on down, that if Total Customer Satisfaction can be attained, the rest of the business takes care of itself. Most operational issues become crises, only as a result of some failure to totally satisfy the internal or external customer. Concurrent with this understanding, which has caused the integration of quality strategy into the day-to-day operations of the business, is a common measurement which directly correlates to customer satisfaction.

LEADERSHIP

At Motorola, the quality culture is pervasive. The CEO formally restated our company objectives, beliefs, goals, and key initiatives in 1987, and quality remained as a central theme. Total Customer Satisfaction is Motorola's fundamental objective. It is this overriding responsibility of everyone in the company and the focus of all of our efforts.

The CEO also reaffirmed two key beliefs that have been part of the Motorola culture since the company began in 1928:

Uncompromising Integrity, and Constant Respect for people.

The CEO has identified three key goals for the corporation:

1. Increased Global Market Share.
2. Best in Class in terms of people, technology, marketing, product, manufacturing, and service.
3. Superior Financial Results.

To achieve these goals and provide Total Customer Satisfaction, Motorola concentrates on five key operational initiatives.

The first of these is Six Sigma Quality. We intend that all products and services are to be at the Six Sigma level by 1992. This means designing products that will accept reasonable variation in component parts, and developing manufacturing processes that will produce minimum variation in the final output product. It also means analyzing all the services we provide, breaking them down into their component parts, and designing systems that will achieve Six Sigma performance. We are taking statistical technologies and making them a part of each and every employee's job, regardless of assignment. Measuring this begins by recording the defects found in every function of our business, then relating them to a product or process by the number of opportunities to fabricate the product or carry out the process. We have converted our yield language to parts per million (ppm), and the Six Sigma goal is 3.4 ppm defect levels across the company. Despite the wide variety of products and services, the corporate goal is the same Six Sigma by 1992.

Our second key initiative, Total Cycle Time Reduction, is closely related to Six Sigma Quality. We define cycle time as the elapsed time from the moment a customer places an order for an existing product to the time we deliver it. In the case of a new product, it is from the time we conceive of the product to the time it ships. We examine the total system, including design, manufacturing, marketing, and administration.

The third initiative, Product and Manufacturing Leadership, also emphasizes the need for product development and manufacturing disciplines to work together in an integrated world.

Our fourth initiative, Profit Improvement, is a long-term, customer-driven approach that shows us where to commit our resources to give customers what they need, thus improving long-term profits. It recognizes that investing in quality today will produce growth in the future.

The final initiative is Participative Management Within, and Cooperation Between Organizations. This approach is designed to achieve more synergy, greater efficiency, and improved quality.

Management demonstrates its leadership in the quality initiative in many ways. The CEO chairs the Operating and Policy committees in twice a quarter, all-day meetings. The Chief Quality Officer of the corporation opens the meetings with an update on key initiatives of the Quality Program. This includes results of management visits to customers, results of Quality System Reviews (QSR's) of major parts of the company, cost of poor quality reports, supplier–Motorola activity, and a review of quality breakthroughs and shortfalls. This is followed by a report by a major business manager on the current status of his/her particular quality initiative. This covers progress against plans, successes, failures, and what he projects to do to close the gap on deficient results, all pointed at achieving Six Sigma capability by 1992. Discussion follows among the leaders concerning all of the above agenda items.

In 1988, Bob Galvin began a more formal program of customer

visits. These visits traditionally had been less systematic and covered only specific topics. Under the new program, members of top management talk to customers at various levels of their business. They ask two basic questions: What do you like about Motorola, and what don't you like? After each visit, a detailed report with specific recommendations is submitted. The visits have become so important that they have been made permanent, and extended to involve all officers. The visits have resulted in the reformulation of our basic goals and objectives mentioned earlier.

Our focus on very specific numeric goals, i.e., 10X, 100X, and Six Sigma capability, is unique in this country.

Quality System Reviews are not a new technology, but we believe very few companies have utilized this system, and it has been successful.

Cycle Time Management is a growing integral part of our programs, and, in conjunction with Six Sigma, represents a very powerful and effective thrust.

Our recognition of quality excellence includes the CEO Quality Award, bestowed on an individual, a team, an operation, or higher level deserving organization by the CEO. This program is administered by the Chief Quality Officer, and the award consists of a plaque, and each individual also receives a uniquely designed pin.

We have applied the principles of Six Sigma Quality to non-manufacturing activities. This gives us a unique ability to benchmark the quality of services. For example, the Communications Sector Engineering Publications Department, using these techniques, reduced defects per equipment manual by a factor of 49 to 1, resulting in a cost savings per manual of 38%.

INFORMATION AND ANALYSIS

In our Six Sigma Quality Programs, the key elements of data are the defects found, compared to the number of opportunities to make defects in the product or process. Throughout Motorola, we have changed our data systems to record defects, opportunities, and the means, variation and limits of both product and process. We direct corrective action through use of Pareto charts, histograms, scatter charts, Ishikawa diagrams, etc.

Analytical techniques begin with the product design cycle, when circuits are analyzed for limit conditions and for Six Sigma distribution of characteristics. Components are evaluated based on internal or supplier data to verify that they will meet the quality and reliability criteria. We make stress and accelerated life tests throughout the engineering design cycle to identify the weak links and to improve product performance by designing in corrections. The environmental stress is always taken well beyond the engineering specifications for the product being designed. Defect per unit goals are established during the design phase of the product and verified through early prototype and pilot runs. Shipping takes place only when the budget is achieved for the defects per unit, and the unit successfully passes an Accelerated Life Test.

Achieving an improvement rate of ten times in two years cannot be done by conventional corrective actions on current products. Such corrective actions must be taken, but they can only yield an evolutionary improvement rate. Thus, the cornerstone of our strategy for dramatic improvement is one that requires each new product to be introduced at a defect per unit level that is typically one-fourth to one-fifth of the product it will replace, or significantly less than similar products in current production.

During the manufacturing phase of products, each step of the process is monitored for its quality performance in defects per million opportunities. Each manufacturing process step is evaluated as either one that can be improved consistent with our goals or as one that must be phased out. For those steps that can be improved, we establish SPC control charts, identify problems and use Pareto charts and histograms to facilitate the needed improvement. As interim goals are reached, tougher goals are set to keep the improvements flowing.

All equipment divisions prepare a management report (the 5-up chart) that shows five graphs of quality performance through the year. The charts show the defects per unit, final quality audit results, out-of-box quality as delivered to the customers via product performance reports, product warranty, and cost of quality for the product or division.

Motorola's Six Sigma Quality program is a unique approach to achieving Total Quality Control (TQC) or Company-Wide Quality Control (CWQC). The Six Sigma Quality program addresses quality in all aspects of the business: products and services, manufacturing and non-manufacturing, administration and operations.

Our Six Sigma Quality program combines the following key ingredients:

1. A superordinate goal of "Total Customer Satisfaction."
2. Common, uniform quality metrics for all areas of the business.
3. Identical improvement rate goals for all areas of the business, based on uniform metrics.
4. Goal-directed incentives for both management and employees.
5. Coordinated training in "why" and "how" to achieve the goal.

STRATEGIC PLANNING

Our goals for both near and long-term improvement all stem from our objective of Total Customer Satisfaction. Our current level for expected results was first delineated in real terms in August 1981. After management elevated one of its senior business managers to become Motorola Director of Quality, it then set about to define the company's new goal going forward. The need for an acceleration of our improvement process had become readily apparent from benchmarking efforts made throughout the company, as well as a realistic projection of where we thought customer expectations would be in five years. In August 1981, the Operating and Policy committees approved our five-year goal to achieve a ten-times improvement in all our efforts on behalf of the customer. These committees represent the highest level of Motorola management.

This quality program was effective and successful. As a result, the Operating and Policy committees tightened the goals in December 1986. The new goals called for quality improvements of 10X in two years, 100X in four years, and achievement of Six Sigma capability in just five years in everything we do. This is one of Motorola's central operational initiatives. We feel these quantitative goals are necessary to achieve Total Customer Satisfaction.

Guided by this broad roadmap of where we would like to be, each business of the corporation incorporates its strategic thinking and plans into its Long Range Plan (LRP). At least once a year the company examines its strategic direction, and all facets of the business, such as new product programs, finances, technology, and quality, on a five-year, or LRP, basis. The first year of such a plan is normally the Annual Plan. Thus, we plan and manage our quality improvement objectives just as we would new product introduction, technology, and any other part of the company's activity. Coupled to these basic quality business goals are goal-directed incentives for both management and employees. These are the Motorola Executive Incentive Plan (MEIP) and the Participative Management Program (PMP). Finally, a large training and education university-like function (Motorola Training and Education Center, MTEC) provides training on quality at all levels, with specific emphasis on providing all employees the knowledge and skills necessary to achieve our quality goals.

We have quantitatively benchmarked best-in-class companies worldwide, and, as a result of these efforts, have driven many of our products and processes to best-in-class levels. Examples include soldering, surface-mounted chip component placement, and cycle time in the production of pagers and cellular mobile radiotelephones. This process allows us to constantly strive to be the best in class in all aspects with true measurements for our goal setting and results.

HUMAN RESOURCES

The mission of Motorola's training function is to provide the right training to the right people at the right time. Our measure of the success of our training program is our having received the American Society for Training and Development's annual award for excellence. Other measures of success can be seen in the scope and internal evaluation of our training.

On the average, we provide one million hours of training per year to our employees. In 1987, we spent $44 million on training. This represented 2.4% of the corporate payroll. Forty percent of the training is devoted to quality improvement processes, principles, technology, and objectives.

During the past years, over 150 hours of quality-related training have been developed, and are being delivered to assembly operators, technicians, engineers, support groups, and the management of these functions. "Course maps" help employees and their managers select programs which respond to individual needs as well as the key corporate initiative of Six Sigma Quality.

Three important parts of our quality training are MTEC training, product/process-specific training, and special management training. Motorola's corporate training function (MTEC) has been able to demonstrate a significant increase in the number of quality-related training programs they have delivered to our employees corporate-wide: In 1985, 37% of MTEC training was devoted to quality; in 1986, 43%

was devoted to quality; in 1987, 73% of the total MTEC training was devoted to quality.

Training programs in Statistical Process Control, Design for Manufacturability, and Understanding Six Sigma are helping us to reach our ambitious goal of 100-fold improvement in four years. These courses provide a set of problem-solving strategies and tools for continuous improvement towards perfection in everything we do. The classes developed by MTEC also provide the framework around which product/process-specific training is added.

To support our commitment to Total Customer Satisfaction through quality in all that we do, we have developed the Motorola Management Institute (MMI). This intensive two-week program in world-class design, manufacturing and quality issues is intended for manufacturing, design, and operational managers at both senior and support levels. Participants build in their professional expertise to enhance their leadership, vision, and decision-making skills. MMI topics include customer-centered culture and marketing for world-class manufacturing and quality, designing for manufacturability, information systems, cycle time management, technology, and supply and change management. Leading experts present the latest information and facilitate the exchange of ideas.

Last year's Senior Executive Program focused on the critical corporate objective of Total Customer Satisfaction. The program involved all 240 officers of the corporation, as well as key customer contact people, such as senior sales representatives and sales/service managers.

In 1970, Motorola formed the Science Advisory Board Associates (SABA). As the organization charter describes it, SABA is dedicated to identifying and rewarding exceptional creative engineering talent and contribution. Approximately 300 Motorola engineers have been honored as associates. In 1988, we inducted twenty-four candidates, including two quality professionals. We had previously inducted six people from Quality Departments over the prior three years.

The crown jewel of recognition awards at Motorola is the Chief Executive Office Quality Award. This prestigious award is presented, with appropriate ceremony, by the corporation's Chairman, Chief Executive Officer, or Chief Operating Office to an individual or group for significant contributions to quality. This award is highly publicized throughout the corporation and is mentioned in Motorola's Annual Report. We've granted forty such awards, involving 5,100 employees, since the inception of the program in 1984.

In 1987, a Chief Executive Office Award was given to produce teams within the Government Electronics Group for providing over 900 space communication systems to the U.S. Government over the past ten years. None of this equipment has experienced a failure. All of the color photographs of the planets sent from Voyager and other deep space probes were sent to earth via Motorola communications equipment.

QUALITY ASSURANCE OF PRODUCTS AND SERVICES

Motorola utilizes various methods to obtain inputs from its customers. The methods vary with the type of input which is being sought, and generally fall within one of the following categories:

Voice of the Customer

Listening to the "Voice of the Customer," the process of assessing the customer's perception of the total quality of Motorola as a supplier, is accomplished by visits to customers by the Chairman, CEO, and other high-level executives within the company. During these visits, customers are made to feel at ease in discussing any area of the business relationship whatsoever, be it product or service. The results of these one-on-one meetings create a top–down driving force within the organization towards the achievement of Total Customer Satisfaction.

Generic Product/Service Requirements

Many new product ideas begin with a unique blend of new technology from the research laboratories, together with customer needs that have been waiting for a solution. There are a number of examples of how Motorola and its customers have combined efforts to bring about a new solution in the marketplace. Some of these are mentioned here:

IBM and Motorola combined efforts to specify, develop, and install a sophisticated nation-wide radio data network to revolutionize the field maintenance and repair of computer systems. Similar technology can now be expanded into the package express and like businesses, based on their unique requirements for data via radio.

Motorola's subsidiary, Computer X, has pioneered cell control and interactive network systems for manufacturing automation, based on the strong needs of the manufacturing commodity (MAP Protocol). Codex worked together with major international airlines in the definition of broad-based data communications networks and the special requirements of the airline industry.

The Bandit paper program of Paging Division is notable, not only for its fully automated manufacturing capability, but for the extension of that automated system to include the order entry process, which in the future will be directly accessible to the customer.

Specific Product/Service Requirements

In the Communications Sector customer requirements for new products are captured in a contract book generated at the beginning of the development effort. The contract contains all features and requirements that the sales organization deems necessary to satisfy customer requirements. In some cases, such as portable two-way radios, blind market surveys are conducted to determine performance, size, weight, cost, and other features preferred by the majority of customers in the marketplace.

The new AIEG Detroit Application and System Engineering Center is an example of Motorola's dedication to enhancement of customer inputs to new product designs. It is an example of Motorola's commitment to the concept of early supplier involvement. Dedicated almost exclusively to engine and vehicle testing on customer systems, it has been welcomed by the customer as a way to jointly specify system related issues which have so far defied formal specification, and to proactively develop solutions to potential problems in future design.

Application Specific Integrated Circuits (ASIC's) are integrated circuits that perform unique functions within a customer's electronic product. SPS has made significant strides in providing design tools which are directly accessible to customers enabling them to perform the otherwise complicated task of designing complex functional devices for a special need. So-called cell libraries are provided by Motorola for use on many of the existing work stations currently in broad use in design laboratories. The next generation of tools, the silicon compiler, will advance the customer's ability to create a device whose function is explicitly defined by the customer to fill a particular need.

RESULTS FROM QUALITY ASSURANCE OF PRODUCTS AND SERVICES

Today Motorola stands along as the only non-Japanese supplier of pagers to Nippon Telegraph and Telephone (NTT). Motorola first earned the right to be a prime supplier in this prestigious Japanese market through the introduction of our highly reliable RC13 pager in 1982. This product was released at a proven reliability level 40% better than the standards then in existence in Japan for communication equipment. Through a program of dedicated design improvements and philosophy of never settling for second best from ourselves or our suppliers, the quality of Motorola paging products, as measured by NTT, has improved seven-fold since this initial product entry. Today, after having shipped more than a half million pagers to Japan, Motorola can be proud of product MTBF figures which typically exceed 130 years. The Motorola Paging Division continues to look toward this Japanese market challenge as a catalyst in our pursuit of product excellence and a symbol of our global competitiveness.

CUSTOMER SATISFACTION

Our customers recognize us as a leader in quality. *Electronic Business* Magazine recently asked 600 companies for information on quality awards they had received in the last two years. The magazine singled out Motorola as receiving close to fifty awards and certified supplier citations. This was the highest number submitted. The number doubled between 1986 and 1987, and the momentum continues.

One basic measure of customer satisfaction is repeat business. Motorola has repeatedly been awarded contracts for the Defense Department's Very High Speed Integrated Circuit (VHSIC) research program, which originated in 1979 and continues today.

We are the only merchant semiconductor manufacturer being funded under Phase 2 of the VHSIC program. Only three of the original six Phase 1 contractors are involved in Phase 2. Under Phase 2, we are developing half-micron technologies, both in CMOS and bipolar. We have supported these operations by installing Class 10 clean rooms in both the Bipolar Technology Center (BTC) in Mesa, Arizona, and the Advanced Product Research and Development Center (APRDL) in Austin, Texas. Also supporting these developments are the Semiconductor Research and Development Laboratory (SRD) in

Phoenix and the former Motorola Integrated Circuit Research Laboratory (MICRL) in Mesa.

Motorola, unlike almost any other *Fortune* 100 company, uses all of its executive officers from both the corporate and sector/group levels to go out, interview and interface with our customers on specific customer satisfaction items. These officers talk with multiple levels of the customer's organization. The results of these interviews are brought back, analyzed, and disseminated throughout all of Motorola. Results are reviewed at Corporate Operating Committee meetings, and follow-up customer meetings continue until issues are resolved and Total Customer Satisfaction is achieved.

B

A Simple, Low-Risk Approach to JIT

Michael J. Hegstad
Hewlett-Packard Company

EXECUTIVE SUMMARY

The movement to JIT manufacturing philosophies and techniques will be critical to the survival of manufacturing entities that compete in the world-wide markets. Unfortunately, the process of converting to JIT can also appear as a survival-threatening experience in itself, but this need not be the case. This paper will describe several simple, low-risk, and low-cost approaches used at the Lake Stevens Instrument Division (LSID) of Hewlett-Packard to successfully achieve a "workorderless" JIT manufacturing environment. In particular, the paper will describe how to gain most of the benefits of JIT in a low-volume, high-mix environment without incurring the complexities and risks associated with a complete post-deduct (backflush) JIT implementation. Three phases of implementation will be described: a pull process with workorders, a workorderless pull process with distinct Stores and WIP inventories, and a partial implementation of a post-deduct process with on-line storage of combined Stores and WIP inventory. The paper will also discuss the risks and trade-offs associated with implementing a post-deduct process and present the results that can be achieved with a less complex strategy for achieving Just In Time manufacturing.

MANUFACTURING ENVIRONMENT

LSID is a high-mix, low-volume, test and measurement instrument manufacturing division with a low level of vertical integration. The division builds thirty-one major products ranging in price from $1,200 to $23,500 and eleven product accessories ranging from $400 to $2,500. The products have long life cycles, and are complex and relatively hard to build, having been designed to achieve state-of-the-art functional requirements rather than manufacturability, in many cases. The products have an average of four options each (and many more combinations), and have build rates varying from 5 to 300 units per month, with most in the 30 to 60 unit per month range.

In addition to the assembly of the products described, the division also produces over 650 subassemblies, which are mainly printed circuit assemblies, sequence kits, and line fab assemblies. A moderate level of automation is used in the PC assembly process where around 20,000 PC assemblies are built per month. Approximately 11,000 active part numbers and 50,000 structure relationships are used in manufacturing LSID's products. There are approximately 400 active vendors for production components. The relatively large number of components and vendors is driven by the broad range of technologies and generations of technology used in the product family.

PHILOSOPHY

At LSID, adopting JIT was interpreted as a continuous process improvement effort focused in the areas of quality, productivity, and flexibility. This philosophy was implemented by focusing on making balanced improvements in quality, production cost, and cycle time, using inventory reduction as the strategy for identifying and prioritizing opportunities for improving these parameters.

Improved quality increases productivity through the reduction of rework, and removes the need for Just-In-Case inventory to work around quality problems. Shorter cycle times and reduced inventories increase process visibility and accelerate the more complete flow of quality information, leading to further process and quality improvements. Increased emphasis on stability and linearity to improve productivity also yields improved quality and reduces the need for buffer inventories.

INVENTORY REDUCTION STRATEGY

Inventory reduction was approached on three fronts: assemblies in WIP, components in WIP, and material in Stores. The first step was to reduce the number of assemblies in process by reducing the size of the queues waiting before each operation. This was accomplished by first shortening MRP lead times, and then by reducing lot sizes. These steps alone had a major impact on reducing cycle times, improving visibility, and simplifying shop-floor tracking and prioritization needs. The next step was to reduce the amount of material in WIP by making smaller, more frequent, and more timely deliveries of material from the stockroom to production. This step squeezed material from the production floor, further improving visibility and reducing space requirements. It also concentrated material back into the stockroom, where it was available for the next replenishment request, which improved service levels by not prematurely dedicating material to certain assemblies. The final step was, and is, to reduce inventory in the stockroom itself through smaller, more frequent deliveries from vendors, and by delivering material directly to production, bypassing Stores. The opportunity and potential gains from these actions are relatively small at LSID, and are only being pursued where reasonable.

The key to making these inventory reductions is to always be working on or delivering the right material, because there is not enough inventory around to squander on something that is not needed immediately. The tools that ensure working on the right priorities are kanban (pull) assembly management and material delivery procedures that only allow inventory to be processed when it has been requested by the next downstream operation. As will be described in more detail later, these two kanban systems are very simple, yet very effective means for managing production and material flow.

ORIGINAL MATERIAL FLOW ENVIRONMENT

The JIT implementation at LSID took place in three phases, corresponding closely to the three inventory reduction strategies described before. Prior to the migration to JIT, a classical MRP/workorder oriented push system was used to control production and material flow. See Figure B.1 for the original material flow process. Material was issued from Stores to WIP using weekly workorders for each product and sub-assembly. The first stop in WIP was in either the PC Check-in or Instrument Kitting operations, where part numbers and quantities were verified, and additional organization (setup and subkitting) was put into the material. The material was then delivered to production. Completed PC assemblies were received into PC fab stock and then issued to the instrument lines to meet with the material coming from instrument kitting. Priorities were driven by MRP due dates and backward scheduled operation due dates from a shop floor control system.

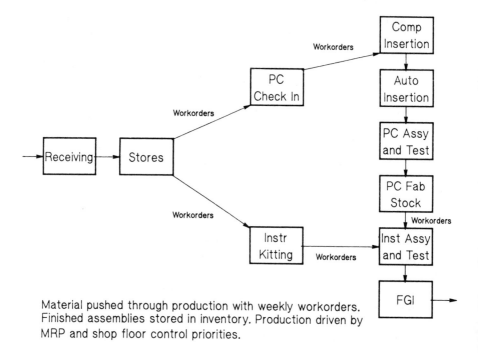

Material pushed through production with weekly workorders. Finished assemblies stored in inventory. Production driven by MRP and shop floor control priorities.

FIGURE B.1 Original environment

IMPLEMENTATION PHASE I: PULL PROCESS WITH WORKORDERS

Phase I of the migration to JIT was very conservative in that no significant changes were made to our information systems. Only procedural changes were made in production and in the Kitting and Check-in areas, but these changes yielded significant results and valuable training. The objectives of phase I were to reduce the number of assemblies in process, and to make workorders invisible to production, even though they would still exist to appease the information systems. See Figure B.2.

One key point: the only way to effectively reduce the risk associated with this type of change is to communicate clearly to everyone what is about to happen and what to expect. Teamwork is essential at all levels. The Material Control organization led the JIT project at LSID, but a manufacturing-wide common vision and common set of

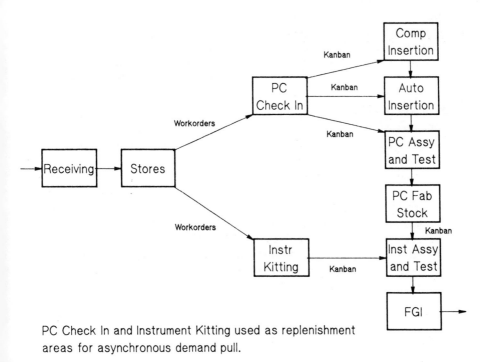

PC Check In and Instrument Kitting used as replenishment areas for asynchronous demand pull.

FIGURE B.2 Phase I process

objectives and measures were the foundation that supported the project. Quality teams, composed of a broad spectrum of people and assignments, solved most of the problems associated with implementing the vision. SQC tools played a major role in the problem-solving processes.

Phase I Material Replenishment

In phase I, material continued to be pushed into the Check-in and Kitting areas using workorders, but the material was then unkitted and stored on shelves in these areas, creating "secondary storage areas" in WIP. Material was no longer delivered to production in workorder organization. In production, a two-bin kanban replenishment process was established where assemblers would request more material by sending an empty kanban box to the secondary storage areas on a part-by-part basis, as needed. Handfuls of material approximating five days of supply were pulled from the WIP on the shelves (no transaction) and delivered back to production using deliver-to-location data printed on the kanban box. Workorders were used to transact material into WIP. To transact completed assemblies back out of WIP, a simple program was written to identify the oldest workorder(s) for the assemblies being completed and then format Receive Workorder transactions to relieve the workorders and update inventory records.

Interesting Observations: Because production was only requesting material as it was needed, the effect was to squeeze a large portion of the components in WIP from the production floor and onto the shelves in the Check-in and Kitting areas. One phenomenon was the existence of backorders to the scheduled workorder pulls in the stockroom when there was plenty of the backordered component on the secondary storage shelves in WIP. When the workorders were turned off going into phase II, the material on the WIP shelves lasted for up to three weeks before material began to be demand-pulled from the main stockroom.

Phase I Assembly Management

To allow production to operate without having workorders to guide them, a kanban (pull) assembly management process was established.

The instrument areas continued to build according to the master schedule, but on a rate per day basis. The building of PC assemblies was driven by a manual kanban card system. When the instrument area had used one kanban's worth of a particular PC assembly, it would simply send a kanban card back to the PC area requesting a kanban quantity of that assembly. Similarly, the PC area sent kanban cards to the Auto-Insertion area to get refills of auto-inserted assemblies. The same was true between axial insertion and component sequencer areas. These procedures replaced the computer-based shop floor control system.

The placement of the kanban signal loops and buffer inventories was determined by the differences in setup cost-driven lot sizes between the major process areas. The instrument lines built in lot sizes of one. The manual PC assembly area began building in lot sizes that would fit in one loading frame (~8–20) and then worked to further reduce lot sizes. The auto-insertion area had more significant setup problems to solve, and started out building quantities about three times as large as the PC assembly area, and then also worked to solve setup problems to enable the use of smaller batch sizes. The component sequencer had major setup problems based on the large number of axial part numbers used and built sequence kits in lot sizes of ten to fifteen days of supply.

To ensure that all assemblies in each production area would proceed through each of the operations at about the same pace, first-in/first-out (FIFO) queue management procedures were implemented. The kanban assembly management procedures allowed production to use smaller lot sizes, and the improved prioritization reduced the need for assemblies in WIP, so MRP lead times were gradually reduced to reflect these changes in the environment. By the end of phase I, production was using what appeared to them to be workorderless kanban assembly management and material replenishment procedures.

IMPLEMENTATION PHASE II: WORKORDERLESS PULL PROCESS

The objective of phase II was to achieve further inventory reductions (and the benefits precipitated) while improving efficiency in the materials areas. See Figure B.3. The strategy was to achieve a workorderless environment throughout manufacturing by modifying and simplifying our information systems and transactions to better match

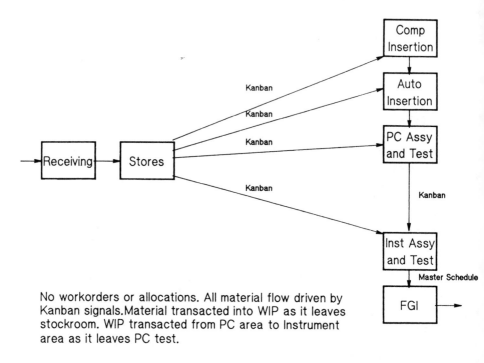

No workorders or allocations. All material flow driven by Kanban signals.Material transacted into WIP as it leaves stockroom. WIP transacted from PC area to Instrument area as it leaves PC test.

FIGURE B.3 Phase II process

the environment we were creating. Phase II was conservative in that we chose to maintain separate and distinct Stores and WIP inventories to provide the visibility to characterize the impacts of our process changes. Phase II was also somewhat conservative in that we chose to modify the ways in which we used our existing information systems rather than implementing new ones. MM3000 in particular displayed excellent flexibility to be molded to fit our needs.

Phase II Material Replenishment

In phase I, workorders existed to transact material into and out of WIP using Issue-Allocation and Receive-WO transactions, respectively. The Check-in and Kitting areas were used to buffer production from these transactions. In Phase II these transactions were replaced by Unplanned-Issue and RTS (Return to Stock) transactions. At this point, classical assembly oriented workorders were no longer needed to transact the flow of inventory. Procedures could then be changed to allow the removal of the Check-in and Kitting areas, returning kanban signals directly to the main stockroom. This had the effect of squeezing the excess WIP components back into the stockroom, where MRP could see them and squeeze them out of the plant.

MRP continued to suggest workorders and used them to compute purchased part requirements, but Production Control simply stopped opening them. MRP sent quite a few expedite requests in response to this lack of action, but the real world operated very well without workorders.

With the Check-in and Kitting areas removed, the material replenishment kanban signals flowed directly back to the main stockroom. When parts were pulled to refill the kanban boxes, they were transacted into WIP using Unplanned-Issue transactions to process workorders to collect cost data. There were about ten of these process workorders, one for each major process area. Completed assemblies were transacted out of WIP using an RTS transaction (Return to Stock for the completed assembly at standard, not the components) from the process workorder. The generation of these transactions was supported by a locally developed stockroom management and kanban replenishment system. When a kanban signal arrives in Stores, a 4-character kanban-ID is bar-code scanned and used to find the corresponding kanban database record. The kanban record contains data such as the part number, the suggested quantity to pull, which process workorder to charge, where to deliver the material, and which process workorder to relieve in the case of assemblies and products.

Advantages of Phase II Procedures

There are two advantages of an Unplanned-Issue based system over workorders and allocations. One is that it provides more flexibility to manage refill quantities and therefore material handling resources. In the kanban system, the number of days that the next refill should last is specified. The kanban system then uses MRP projected usage data to convert that Days of Supply (DOS) number into a requested quantity. By setting the refill DOS lower for large, expensive, and easily damaged parts, and higher for small, cheap, and hardy parts, material handling resources can be better leveraged than by issuing matched sets of parts to workorders.

The other advantage is that unplanned issues allow pulling parts in convenient (often prepackaged) quantities, which increases both material handling efficiency and inventory accuracy, while decreasing material handling related damage, like electrostatic discharge (ESD). In the kanban system, the requested quantity is only a suggestion. The operator may over- or underissue the requested quantity by up to 50%. If underissuing, the operator must create a backorder if there are not enough parts to meet the request, but a backorder is not created if the underissue was only a matter of convenience. The system is robust and self-correcting. Overfilled kanbans will take a little longer to cycle through and underfilled ones will return a little sooner.

A major feature of the phase II procedures is the separation of the Stores and WIP inventories. Very accurate on-hand balances can be maintained in Stores for procurement planning purposes, and verified using cycle-counting procedures. WIP inventory can be estimated using information from the kanban database, and is verified quarterly by taking a physical inventory of WIP. Any scrap or other unexpected usage of material is also automatically captured using these procedures. Unexpected usage simply causes the kanban container to need refilling sooner, which causes the Stores on-hand balance to be decremented, which notifies MRP and the buyer of the increased usage, providing for timely response.

One additional change was made to the physical material handling procedures during phase II. Some of the production areas changed from a 2-bin to a 1-bin kanban system. Instead of waiting for the first bin to become empty and then working out of the second bin while the first was being refilled, the assemblers use only one bin and send only the kanban label back to Stores when there are just enough parts left to last until the replenishments arrive. This process enhancement cuts the number of components in WIP by approximately 50% without increasing the material handling workload or transaction volumes.

Disadvantages of Phase II Procedures

The only major potential disadvantage of the phase II procedures is the effect that the loss of workorders has on MRP for procurement planning. Workorders provide MRP visibility of the assemblies and components in WIP. Without them, MRP assumes that there is nothing in process to meet the demands of the master schedule, and tries to rapidly replenish the material that is physically already in WIP, but invisible to MRP. If this effect is not compensated for, MRP will overplan requirements and bring material into the facility ahead of when it is actually needed. The solution chosen is to reduce the assembly lead times in MRP to compensate for the number of days of supply of inventory in WIP. For example, if the actual time to build an assembly is six days, and there are four days of supply of inventory in WIP, then the assembly lead time should be set to two days. Whenever the actual build time is less than the days of supply in WIP, the assembly lead time is set to zero. This compensation causes MRP to schedule the receipt of material for its planned workorders after the building of the assemblies actually begins, which is okay because the assembly will be built from the invisible material already in WIP.

The MRP system continues to be a reasonably close approximation of the actual environment, but it does not have the resolution of an MRP system supported by workorders and allocations. An allocationless environment also induces some MRP nervousness as material leaves the stockroom in different quantities and intervals than planned. This effect is minimal, however, if the WIP replenishment quantities are small relative to the supplier replenishment quantities.

Phase II Assembly Management

Phase II assembly management continued to operate as described in phase I. Progress continued on solving many of the setup problems which allowed lot sizes and buffer inventories to be reduced. Enough of the setup problems in auto-insertion were solved to allow the auto-insertion area to build in the same lot sizes as the manual PC assembly and test area, which allowed the removal of the kanban loop and buffer inventory between PC and AI. Most assemblies are now built in quantities of three or four, which represents approximately 0.3 to 1.0 day of supply.

IMPLEMENTATION PHASE III: PULL PROCESS WITH POST-DEDUCT AND ON-LINE STORES

The objective of phase III is to achieve additional efficiencies and inventory reductions by removing the distinction between Stores and WIP inventory and allowing material to be delivered directly from Receiving to production, bypassing Stores. The strategy is to convert

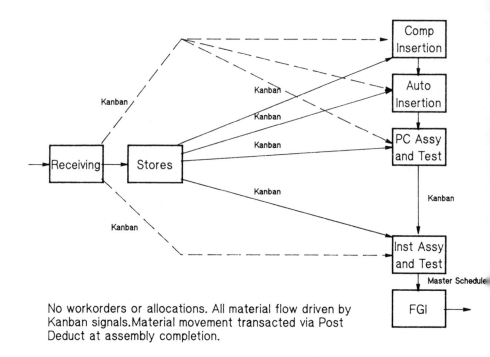

No workorders or allocations. All material flow driven by
Kanban signals.Material movement transacted via Post
Deduct at assembly completion.

FIGURE B.4 Phase III process

all components in WIP to Stores inventory by simply not transacting
material into WIP and using post-deduct (backflush) transactions to
relieve inventory as assemblies are completed. For some parts and
processes, secondary stockrooms were established in production and
Move transactions used to transact material from the main stockroom
to the secondary locations. A specialized form of this process has
been implemented in the sequencer area, and investigation for further
implementation is continuing. See Figure B.4.

Phase III Material Replenishment

In phase II, all material passed through the stockroom before being
issued into WIP, but in phase III a portion of the material could be
delivered directly from Receiving to the point of use in production.
Some of the material replenished from the main stockroom is trans-

acted to secondary stockrooms using Move transactions when additional inventory visibility and control are desired. On-hand balances for some components are decremented as assemblies are completed. Completed assemblies continue to be RTS'd from the process workorders.

Potential Advantages and Disadvantages of Phase III Procedures

Direct material delivery from Receiving to production, bypassing Stores, removes unnecessary material handling steps and reduces storage space in the stockroom. It also makes the components in WIP visible to MRP again to improve procurement planning resolution. Unfortunately, these gains come at the cost of increased risk, complexity, and support costs. In a low-volume, high-mix environment the benefits do not appear to justify the cost when compared to the phase II processes. High-volume, low-mix producers are likely to draw different conclusions based on their ability to leverage more heavily from direct delivery and space savings, but we feel that only certain commodities (ICs and blank PC boards) warrant the implementation of phase III procedures, and even there the net gains are relatively small. The reasoning behind this conclusion is detailed in the following.

Some of the reasons that phase III is not attractive to LSID are factors of the technological and business environment. Should these factors change, as outlined in the following, phase III could become more appealing.

1. Reduction of the vendor base and selection of vendors located geographically close, to make frequent deliveries of small quantities economically feasible.
2. Component part standardization across products to increase usage volumes to make frequent deliveries cost-effective, and to reduce storage space requirements in production.
3. Further progress towards generic production lines, combined with a material presentation strategy that would minimize the number of locations for a component in production, to minimize storage space requirements and inventory management resources.
4. Development of vendors and vendor processes that can react rapidly to changes in demand.

Phase III Concerns and Trade-Offs

Perhaps the largest concern is that post-deduct procedures are biased towards generating material shortages and production disruptions. Using post-deduct processes, MRP assumes that material is available a little longer than it really is. More seriously, any material usage not transacted by post-deduct or scrap transactions results in on-hand balances that overstate the amount of material available, which causes MRP to underplan or delay replenishment. Phase II procedures, on the other hand, are biased towards providing a small layer of safety stock due to the invisible WIP and are robust in terms of responding to untransacted usage. Any unexpected usage is reflected by the next kanban refill transaction which decrements the on-hand balance, causing MRP to recognize the increased usage and plan correctly.

As shown before, post-deduct systems demand very accurate and timely product structure data because this data is used to manage on-hand balances. Any structure errors will result in corrupt inventory and procurement planning data. Production change orders (PCO's) become much more difficult to implement because the structures must be updated exactly when the first assembly containing the new material is to be post-deducted. In contrast, phase II procedures use only simple transactions for maintaining inventory integrity. MRP still requires accurate product structure data, but the phase II environment is not as sensitive to errors, and the timing of PCO data updates is much less critical.

Another issue is that material transaction volumes increase dramatically in a post-deduct environment. There are Receive transactions for all parts, Move transactions between primary and secondary stockrooms for many parts, RTS transactions for completed assemblies, and Issue transactions for each component in an assembly each time the assembly is post-deducted. In phase II there are only Receive, Issue, and RTS transactions, and the Issue transactions are usually for much larger quantities than the post-deduct issues. It is estimated that post-deduct procedures would increase transaction volumes by at least 700%.

A clear post-deduct advantage is that it provides MRP visibility of all material for procurement planning purposes. This would be a major advantage if there were large amounts of invisible WIP, and if the MRP overplanning compensation techniques did not work well. But in the current phase II environment, the quality of the procurement planning data is more than adequate. In fact, there are concerns that the data dependencies, vulnerabilities, and biases described before would actually lead to less dependable procurement planning information.

Finally, the material handling efficiencies provided by post-deduct are at least offset by the additional inventory management required. Cycle counting is very difficult in an environment where there is no distinction between Stores and WIP inventory, and on-hand balances are not decremented until some time after the parts have physically been removed from their storage location. Material storage in production is also less efficient than in the stockroom. Moving inventory from the main stockroom to on-line stores in production results in an overall increase in floor space needs.

SUMMARY

Large and rapid material flow and manufacturing process improvements can be realized from several safe and simple changes to material management procedures and transactions. It is suggested that even those entities that can benefit from post-deduct procedures take a phased implementation approach to reduce risks and start receiving significant benefits early on in the migration process. It is also suggested that those entities currently pursuing a post-deduct environment question whether the benefits outweigh the associated complexity, risk, and costs, especially when compared to the alternative procedures described in phase II before.

The metrics of Table B.1 were chosen as the key indicators of progress towards JIT manufacturing. The phase II+ results primarily reflect the fine tuning of phase II processes.

Cycle time is a key metric because it indicates the ability to respond to changes in customer demand and product mix. The cycle time metrics are volume weighted averages for all products and assemblies built in the division. Manufacturing Cycle Time measures the elapsed time from the beginning of the auto-insertion process to the delivery of a complete and tested product to Finished Goods Inventory (FGI). PC Cycle Time measures the elapsed time from the beginning of the auto-insertion process to the delivery of a tested PC assembly to the instrument lines. In both metrics, component sequencing is treated as an off-line activity and not included in the metric. Stores Cycle Time measures the elapsed time from the receipt of

TABLE B.1 Results at the End of Implementation of Each Phase

	Original (Nov. 1984)	Phase I (Jan. 1986)	Phase II (Jan. 1987)	Phase II+ (April 1988)
Mfg Cycle Time (Days)	47	28	9	5.4
PC Cycle Time (Days)	17	9	3	1.4
Stores Cycle Time (Hours)	43	43	4	1.6
WIP Inventory ($K)	6,600	5,080	3,995	3,050
Fab Inventory ($K)	1,347	1,278	390	370
Stores Inventory ($K)	12,200	10,675	8,164	7,255
Stores Occupancy (Sq. Ft)	23,800	22,700	17,390	16,060
Material Control Staff	67	55	31	23
Stores Txns/Person/Day	55	88	106	126

a signal to pull material (the pull deck in the workorder environment) until the material is delivered to production.

Inventory levels are monitored because inventory reductions are used as the catalyst to drive process improvements. Occupancy is measured to ensure that we are taking advantage of the space made available by inventory reductions and process improvements. Many space savings occurred in production, but they were difficult to measure accurately due to other factors, such as new product introductions and changes in production volumes, so Stores Occupancy became the main indicator of space utilization. Staffing levels and productivity metrics are monitored because they indicate the results of process improvements. Again, there were JIT related productivity improvements and people savings in many areas, but the changes in material control were the most measurable and attributable to JIT. Examples of savings outside of Material Control include expediters, cost accountants, buyers, production supervisors, and direct labor people.

C

The Role of Product Design in Competitive Manufacturing*

Kenneth A. Crow

INTRODUCTION

Competitive manufacturing requires improved product quality and the elimination of waste in manufacturing. The process of achieving these objectives begins when the designer conceptualizes a product to meet a market or a customer need. The design process, through utilization of Design for Manufacture (DFM) concepts, should focus on developing a product that both meets customer needs and can be effectively manufactured. In this way, product and process design will be a key factor in achieving competitive advantage in manufacturing.

In the broadest sense, quality is defined as providing a product

---

*This paper is based on material originally presented at the Second International Conference on Design for Manufacturability, an annual event sponsored by Management Roundtable, Inc., 1050 Commonwealth Avenue, Boston, MA 02215, (617) 232-8080.

and related services to satisfy the customer. This satisfaction is represented in a number of dimensions:

- Product performance and capabilities
- Product durability and reliability
- Product styling and aesthetics
- Perceived quality of the product and customer emotional satisfaction
- Product maintainability and after-sale service and support

These dimensions of satisfaction primarily relate to the design of the product. In other words, quality begins with the design of the product.

In the past, products have been designed that could not even be produced. Products have been released for production that could only be made to work in the model shop when prototypes were built and adjusted by highly skilled technicians.

To be competitive and better satisfy the customer, the process must begin by knowing what the customer wants, designing those requirements into the product, and then ensuring that both the factory and the virtual factory (the company's suppliers) have the capability to effectively produce the product. This is a facet of the two dimensions of "Integrated Manufacturing":

- Functional integration whereby product and process design are accomplished in a way to optimize:

  Manufacturing planning and control

  The production process

  Distribution

  After-sales service and support

- Logistical and market integration to consider capabilities, needs, and timing of materials and products to meet ultimate customer demands.

PRODUCT DESIGN OBJECTIVES

Products are initially conceptualized to provide a particular capability and meet identified performance objectives and specifications. Given these specifications, a product can be designed in many different ways.

The designer's objective must be to optimize the product design with the production system. The production system includes suppliers, material-handling systems, manufacturing processes, labor-force capabilities, and distribution systems.

Generally, the designer works within the context of an existing production system that can only be minimally modified. However, in some cases, the production system will be designed or redesigned in conjunction with the design of the product. When Design Engineers work together with Manufacturing Engineers to jointly design and rationalize both the product and processes, it is known as "integrated product development," "simultaneous engineering," or "concurrent product and process design." The term "design for manufacture" (DFM) describes the process where designers and engineers consider the enterprise's total production system and design a product that can be effectively manufactured with this production system. In considering the total production system, the designer will have a higher likelihood of meeting customer needs in terms of the three dimensions of manufacturing performance: quality, cost, and schedule.

The main goals of DFM are to:

- Improve product quality
- Increase productivity and capital utilization (e.g., labor, inventory, and fixed assets)
- Reduce leadtime—both the leadtime to bring a product to market as well as the recurring manufacturing leadtime
- Maintain flexibility to adapt to future market conditions

DFM is intended to prevent product designs that:

- Simplify assembly operations at the expense of more complex and expensive component fabrication
- Simplify component fabrication at the expense of complicating the assembly process
- Optimize production in one specialized factory while increasing material acquisition and product distribution costs
- Result in the simple and inexpensive production of a product that is difficult and expensive to service and support

A designer's primary objective is to design a functioning product within given economic and schedule constraints. However, research has shown that decisions made during the design period determine 70% of the product's cost while decisions made during production only account for 20% of the product's costs.

The application of DFM must consider the design economics. It must balance the effort and cost associated with refinement of the design to the cost and quality leverages that can be achieved. In other words, greater effort to optimize a product design can be justified with higher value or higher volume products.

Therefore, if an organization is to rationalize the product design and the production system to minimize costs, improve quality, reduce leadtime, etc., it must accomplish the following:

- Identify DFM design rules
- Utilize engineering tools to develop and evaluate design alternatives
- Manage product and process design in an integrated way
- Train engineering and design personnel to achieve these objectives

PRODUCT DESIGN RULES

A number of design and manufacturing guidelines and principles have been established over time to achieve higher quality and lower cost products, greater market responsiveness, improved assembly, productivity, improved application of factory automation, and better product reliability and maintainability.

In addition to these guidelines, designers need to understand more about their own company's production system—its capabilities and limitations—in order to establish company-specific design rules. These rules will be used to further guide and optimize their product design. For example, designers need to understand the tolerance limitations of certain manufacturing processes.

The following DFM guidelines provide an example of how to achieve improved effectiveness with product designs:

1. *Reduce the number of parts* because for each part, there is an opportunity for a defective part and an assembly error. The probability of a perfect product goes down exponentially as the number of parts increases. As the number of parts goes up, the total cost of fabricating and assembling the products goes up. Automation becomes more difficult and more expensive when more

parts are handled and processed. Costs related to purchasing, stocking, and servicing also go down as the number of parts are reduced. Inventory and work-in-process levels will go down with fewer parts. As the product structure is simplified, fewer fabrication and assembly steps are required, manufacturing processes can be integrated, and leadtimes further reduced. The designer should go through the assembly process part by part and evaluate whether the part can be eliminated, combined with another part, or the function can be performed in another way. To determine the theoretical minimum number of parts, ask the following: Does the part move relative to all other moving parts? Must the part absolutely be of a different material from the other parts? Must the part be different to allow possible dis-assembly?

2. *Foolproof the assembly design* (poka-yoke) so that the assembly process is unambiguous. Components should be designed so that they can only be assembled in one way; they cannot be reversed. Notches, asymmetrical holes, and stops can be used to foolproof the assembly process.

3. *Design verifiability into the product* and its components. For mechanical products, verifiability can be achieved with simple go/no-go tools in the form of notches or natural stopping points. Products should be designed to avoid or simplify adjustments. Electronic products can be designed to contain self-test and/or diagnostic capabilities. Of course, the additional cost of building in diagnostics must be weighed against the advantages.

4. *Avoid tight tolerances* beyond the natural capability of the manufacturing processes. Otherwise, this will require that parts be screened for acceptability. Also, avoid tight tolerances on multiple, connected parts. Tolerances on connected parts will "stack-up," making maintenance of overall product tolerance difficult. Design in the center of a component's parameter range to improve reliability and limit the range of variance around the parameter objective.

5. *Design "robustness" into products* to compensate for uncertainty in the product's manufacturing, testing, and use. "Robustness" is excessive capability to ensure quality performance. While a general quality goal is to reduce variance within the production system, "noise" exists outside the system which cannot be controlled. Robustness is intended to offset this external noise as well as uncontrolled variances within the production system. An ex-

ample of robustness to deal with uncertainty in the manufacturing process is a RAM semiconductor chip where extra cells are designed on the chip and can be activated to compensate for defective cells. Robustness in mechanical parts can be achieved by strengthening the part with added material or stronger materials to compensate for uncertainty in loading, torque, and damage.

6. *Design for parts orientation and handling* to minimize non-value-added manual effort and ambiguity in orienting and merging parts. Parts must be designed to consistently orient themselves when fed into a process. Product design must avoid parts which can become tangled, wedged, or disoriented. Part design should incorporate symmetry, low centers of gravity, easily identifiable features, guide surfaces, and points for pick-up and handling. This type of design will allow the use of automation in parts handling and assembly such as vibratory bowls, tubes, magazines, pick and place robots, vision systems, etc. When purchasing components, consider acquiring materials already oriented in magazines, bands, tape, or strips.

7. *Design for ease of assembly* by utilizing simple patterns of movement and minimizing fastening steps. Complex orientation and assembly movements in various directions should be avoided. Products should be designed which snap together or can be readily bonded. Assembly with nuts, bolts, and washers should be avoided. Also avoid electrical cables and other flimsy parts. Their flexibility makes material handling and automation difficult. Avoidance of electrical cables will also avoid wiring errors, which are a major source of quality deficiencies. Plugs and connectors are a preferred alternative. The product's design should enable assembly to begin with a base component upon which other parts are added. Assembly should proceed vertically with other parts added on top and positioned with the aid of gravity. A product that is easy to assemble manually will be easily assembled with automation. Assembly that is automated will be more uniform, more reliable, and of a higher quality.

8. *Utilize common parts and materials* to facilitate design activities, to minimize the amount of inventory in the system, and to standardize handling and assembly operations. Common parts will result in lower inventories, reduced costs, and higher quality. Operator learning is simplified, and there is a greater opportunity for automation as the result of higher production volumes and

operation standardization. Limit exotic or unique components because suppliers are less likely to compete on quality or cost for these components. Group technology (GT) can be utilized by designers to facilitate retrieval of similar designs and material catalogs, or approved parts lists can serve as references for common purchased and stocked parts. GT can also be used to guide in the development of manufacturing cells for common part or product families, thereby minimizing inventory and providing improved effectiveness through manufacturing focus.

9. *Design modular products* to facilitate assembly with building block components and sub-assemblies. This modular or building block design should minimize the number of part or assembly variants early in the manufacturing process while allowing for greater product variation late in the process during final assembly. This approach minimizes the total number of items to be manufactured, thereby reducing inventory and improving quality. Modules can be manufactured and tested before final assembly. The short final assembly leadtime can result in a wide variety of products being made to a customer's order in a short period of time without having to stock a significant level of inventory. Production of standard modules can be leveled and repetitive schedules established.

10. *Design for ease of servicing* the product. Easy access should be provided to parts which can fail or may need replacement. These parts cannot be permanently attached so the designer should consider the use of quarter-turn screws, latches, twist locks, threaded parts, and spring clips. The design should be modular with easily disconnected modules for replacement. When designing for ease of servicing, there needs to be consideration of the trade-offs involved in providing for ease of repair. In high-reliability and low-cost products, designing for ease of repair is not worthwhile. In the case of a product with parts subject to wear, maintainability may be more important than initial product acquisition cost, and the product must be designed for ease of repair.

EVALUATION OF DESIGN ALTERNATIVES

With the traditional design approach, the designer would develop an initial concept and translate that into a product design, making minor modifications as required to meet the specification. DFM requires that the designer start the process by considering various design concept alternatives early in the process. Only through consideration of more than one alternative is there any assurance of moving toward an optimum design.

Using some of the previous design rules as a framework, the designer needs to creatively develop design alternatives. Then alternatives are evaluated against DFM objectives.

A variety of engineering tools are available to assist in the economic development of multiple design alternatives as well as the evaluation of these alternatives. These design tools include computer-aided design (CAD), computer-aided engineering (CAE), solids modeling, finite element analysis, group technology (GT), and computer-aided process planning.

CAD/CAE aids the designer in cost effectively developing and analyzing design alternatives. Solids modeling helps the designer visualize the individual parts; understand part relationships, orientation and clearances during assembly; and detect errors and assembly difficulties. Finite element analysis and other design analysis tools can be used to assess the ability of the design to meet functional requirements prior to manufacture as well as assess a part's or product's robustness. Group technology coding and classification capabilities can assist in the retrieval of similar designs and avoid redesign of additional parts. Computer-aided process planning can work with CAD and GT data to generate process plans for manufacture and assembly of products, thereby helping the designer assess the manufacturability of a design.

However, the use of these design productivity tools must be managed because they create a temptation for the designer to exercise too much creativity and design a slightly improved part rather than opt for part standardization.

In addition to these design productivity tools, there is a variety of DFM analysis tools to evaluate designs and suggest opportunities for improvement. One is the Boothroyd and Dewhurst Analysis Program. This analyzes design symmetry; ease of part handling, feeding, and orientation; and the number of parts. Other programs are available to analyze assembly operations, evaluate designs against design practices, and analyze tolerancing requirements.

In summary, these tools should be used to:

· Allow the designer to develop design alternatives economically
· Evaluate these alternatives against DFM objectives

- Aid in understanding DFM and how to apply it
- Establish standardization designs based on DFM principles which can be readily retrieved for new products

INSTITUTIONALIZING DFM

DFM can only be successfully applied when there is a management commitment to make it happen. The process must start with improved communications between engineering and manufacturing.

This communication begins with the two organizations working together to define design rules. Engineering must also understand manufacturing processes and operating constraints as a perspective on the design rules and process limitations. This communications process can be facilitated with cross-assignments so that personnel in each group gain an appreciation of the broader picture of the company's operations. Co-location of design and manufacturing engineering personnel is another step. Finally, integrated product and process design teams is a further step to improve design effectiveness.

The designer's understanding of DFM and integrated product and process design in his own company can be facilitated with education and training covering the following subject areas:

- Materials and processes
- CAD/CAE/CAM
- Manufacturability guidelines
- Design for manufacture and DFM examples
- Design assessment tools
- Economic assessment

Once the designer acquires a basic DFM background, he must learn to work closely with manufacturing engineers and others who can provide him with feedback on DFM design issues. This working relationship is also formalized in periodic design reviews. For the design review to be effective, it must:

- Be formalized
- Be held at points when feedback can positively influence the design process (not too late in the process)

• Utilize a set of review guidelines
• Be conducted in a positive environment of cooperation

Suppliers should also be involved in the integrated design and the design review process. No one knows more about the materials or components to use in a product than the supplier of those items.

As stated earlier, the optimum is for the design of both the product and the process to proceed in parallel as a tightly integrated activity.

While DFM will require some additional effort on the part of the designer early in the process, sharing information between manufacturing and engineering will result in a quicker and smoother transition to manufacturing and a lower total cost. This impact can be partially seen in a number of studies that have shown the debilitating effects that engineering change orders (ECO's) have on manufacturing productivity. When ECO's can be minimized through early consideration of the impact of a design on manufacturing effectiveness with DFM, the initial effort to implement DFM can be more than justified.

SUMMARY

DFM is a critical part of achieving competitive advantage in manufacturing since such a significant part of a product's cost, manufacturing leadtime, and quality is a result of its design.

Product simplification and parts reduction can have a multiplier effect on reducing inventory, leadtime, and manufacturing cost. If the design process is accomplished within a framework of group technology and parts standardization, a further reduction in inventory, leadtime, and manufacturing cost can be achieved.

A reduced number of parts, parts standardization, and the resulting higher parts volume facilitates:

• Automation and resulting process consistency and cost reduction
• Leveled and repetitive production schedules
• Development of manufacturing cells
• Reduced chance for errors in fabrication and assembly
• Better integration of design with manufacturing process capabilities, thereby improving product quality

By developing product designs based on modular building block components, additional results like those given are achieved. A wide variety of products can be assembled from a relatively small number of building blocks. Further, this approach can result in a short final assembly process, enabling products to be built to order more competitively.

Many organizations are finding that DFM practices are allowing them to make significant improvements in their competitiveness without major investments in plant and equipment. However, effective utilization of DFM and integrated product and process design will require that organizational and cultural barriers be broken down. Engineering and manufacturing personnel must find ways to effectively work as an integrated team during product and process design. This organizational and business practice change is a challenge for many companies.

As more companies implement and utilize the philosophies and tools of just-in-time (JIT) production, total quality control (TQC), and computer-integrated manufacturing (CIM), the role of product design will become crucial. Product design may be the ultimate way to distinguish a company's capabilities. It will then become critical to achieve competitive advantage through the development of high-quality, highly functional products effectively manufactured through the synergy of integrated product and process design.

D

An Automated DFM Analysis System for PC Board Assemblies

CLARK NICHOLSON and JERRY STONE
Lake Stevens Instrument Division, Hewlett-Packard Company

This paper describes a system that analyzes printed circuit board designs for manufacturability. The system consists of a relational database containing component and design information, and a series of programs that analyze this data, assist in component selection, generate process plans for the designs, and suggest ways of changing designs to improve their manufacturability.

Since this system has been implemented, improvements in cost, quality, and cycle time have been realized. Both quantitative and qualitative results are discussed. The key conclusions presented in this paper are: (1) automated DFM analysis is much more timely and thorough than manual analysis; (2) process planning is the first step in DFM analysis, therefore, DFM analysis should be an integral part of a computer integrated manufacturing (CIM) system; and (3) flexibility to handle different processes is very important when implementing a DFM or CIM system.

INTRODUCTION

For the past several years Hewlett-Packard's Lake Stevens Instrument Division (LSID) has been focusing on improving cost, quality, and cycle time to gain a competitive advantage in the market place. This is best achieved by affecting the manufacturability of a product as early in the design cycle as possible. This is referred to as design for manufacturability (DFM) and can be defined as the process of designing products that have lower assembly cost, higher quality, and reduced cycle times. Since the majority of the value added to the instruments assembled at LSID is in the printed circuit (PC) board assemblies, we have concentrated our DFM efforts on PC board designs.

How does one design a PC board for manufacturability? Figure D.1 shows that auto-insertion plays a major role in the cost and quality of a PC board assembly in manufacturing. As the percentage of auto-insertable components increases, the cost per insertion and the number of defects are reduced.

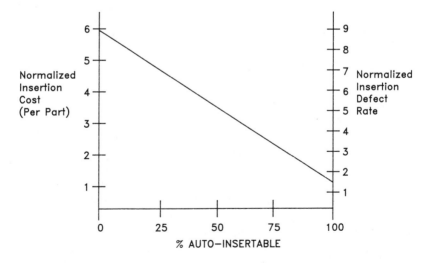

FIGURE D.1 Normalized insertion cost and defect rate vs. auto-insertability

LSID designs and manufactures spectrum and network analysis instrumentation in the 200-MHz frequency range. The PC board assemblies found in these products contain a wide variety of analog and digital components, including axial-leaded, radial-leaded, dual in-line packages, surface mount, and various odd-shaped devices. Due to the dense population and wide variety of components used in this type of PC board, achieving a high rate of auto-insertability is very difficult.

The first DFM project at LSID was a book of detailed design rules for producing a PC board of minimum cost and maximum quality. Using these guidelines, Process Engineers invested many tedious hours manually analyzing PC board drawings and suggesting changes to improve manufacturability. Today, a DFM analysis system has been developed to automate this process, reducing analysis errors and increasing the productivity of process engineering. This system helps designers to create more manufacturable PC boards.

AUTOMATED DFM ANALYSIS SYSTEM OVERVIEW

The PC board design process consists of three major steps: (1) component selection; (2) schematic entry, which defines the material list and the electrical connections between the components; and (3) PC layout, which defines the physical connections between the components. The automated DFM analysis system attempts to help the designer make the correct decisions at each of these steps. Figure D.2 is a block diagram of this system showing the major pieces and how they interact.

This system consists of computer aided design (CAD) tools, a relational database containing component and assembly data, and a series of programs that access and analyze this data. The CAD tools are used to enter schematic and PC board design information. The relational database contains much of this design information (in a more accessible format) plus component and process information.

COMPONENT ANALYSIS

The component selection process is facilitated through the use of the component information in the relational database and a user interface. This is an on-line electronic value sort which enables the designer to search for components based on electrical and/or mechanical param-

238

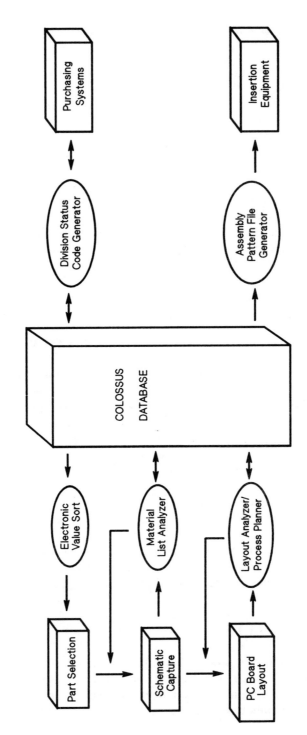

FIGURE D.2 Automated DFM analysis block diagram

eters. As the first step in the DFM system, a program assigns a division status code to each component in the database. The division status code is a manufacturability rating reflecting division specific manufacturing process and procurement information.

The division status code is composed of individual ratings (each from 1 to 5) from the division Purchasing, PC Assembly, and Component Engineering Departments. If any of the three individual ratings are code 5 (unacceptable), the overall rating is set to code 5. Otherwise, the division status code is computed by simple average of the individual divisional ratings. The final division status code is in the range of 1.0 to 5.0 and appears in a two-character field on the value sort screens as 10 to 50.

The Purchasing rating is based on quantitative performance evaluations of each possible supplier of a component. These evaluations reflect the supplier's past performance in technology, quality, responsiveness, delivery, and cost. The PC Assembly rating is based on the component's cost per insertion. A "component insertability analysis" program generates a list of all possible insertion areas for the component using division specific process data and the component's package and termination data. The insertion area with the lowest cost per insertion defines the PC Assembly rating for the component.

The Component Engineering rating provides divisional input on part reliability and technology issues. This rating is implemented on a party-by-part basis as an override to the HP corporate component rating.

In addition to the division status code, other critical divisional purchasing and assembly information are available on a divisional data screen. An example of this screen is provided in Figure D.3. The individual department ratings are shown to provide insight into how the aggregate division status code was generated.

The divisional recommended supplier is displayed to ensure the designer uses the vendor with the best rating. The assembly process and cost per insertion are shown for reference. In addition, the lab stock location is shown so the component can be located quickly and easily. The division status code and related divisional data are assigned for all components in the database, even those which are not currently used or purchased at LSID.

```
SEARCH Menu:  QUERY  Browse  CPIR  DIVD  MatL  HELP  GoBack  Exit
+----------------------------------------------------------------------------+
|Div_Cos Menu:  NEXT  PREVIOUS  FILEOUT  OUTPUT_PRINT  GoBack  HELP  Exit
| Press (RETURN) or (NEXT) or (N) to display next page.                      |
|                                                                            |
|               PART INFORMATION REPORT - Division Cost/Usage        Total 29 |
| Part number 0764-0016..................................Page  1 of  6
| List Price          Present Price        Target Price       Trend          |
| $0.00               $0.07                $0.00
| Contract                   Stock Location      Buyer  6MoReq     Insp  Lead     Type |
| C                                                71    2134        C    30         2
| Comments                                                                   |
| BRADFORD ELECTRONICS INC                | AUTO-INSERT (VCD AXIAL) $0.0044/ins
| Units      Class  ABC  LSFlag  LS_Loc  LS_Eq  DivStat  DivFlag  Date       |
| EA           P     D       Y    8-F             13         Y    12/20/1988
| Division Status Comments                                                   |
| Insertion: 1, Purchasing: 2, Component Engr: 1
| Div* Kdx  Yr Usage  UM   Std-Cost   Dom-Cost   Int-Cost Div_Stat P  V  S  A  B |
| 04   01        520  EA   $    .09  $     .11  $     .11   1  L   N  Y  N  N  N
| 08   01        860  EA   $    .08  $     .10  $     .10   *  *   N  N  N  N  N
| 09   01       6136  EA   $    .07  $     .08  $     .08   1      N  N  N  N  N
| 10   05       1226  EA   $    .07  $     .08  $     .08   1  L   N  N  N  N  N
| 11   5              EA   $    .11  $     .13  $     .13   *  *   N  Y  N  N  N
+----------------------------------------------------------------------------
```

FIGURE D.3 Divisional data screen example

MATERIAL LIST ANALYSIS

Once a schematic has been generated for a design, using a CAD tool, the material list is loaded into the database. These data are then evaluated by the "material list analyzer," and an initial process level plan (Nicholson, 1988) is generated. Process level planning consists of assigning an insertion area to each component in a design. This is done by a rule-based (Winston, 1984) analysis of the possible insertion areas for each component. (The possible insertion areas were determined by the component insertion analysis program, described in the previous section.)

This initial process plan is automatically analyzed, and a component analysis report is generated, suggesting ways to make the PC board design have a lower assembly cost and higher quality. All components that cannot be inserted in a preferred insertion area or are not a preferred part are reported in the component analysis report. The material list analyzer searches the component database in an attempt to find equivalent components that can be substituted to improve the manufacturability of the PC board. The designer may take these suggestions, make changes to the schematic, and rerun the material list analyzer. This may be repeated several times until the designer is satisfied that the material list is as manufacturable as possible.

LAYOUT ANALYSIS

After the schematic has been completed, a PC board layout is designed. The PC board layout defines the physical locations and connections between the components and the blank board outline. The blank PC board information and the component locations are then loaded into the database. These data are evaluated by the "layout analyzer," and a final process level plan and machine level plan (Nicholson, 1988) are generated.

The final process level plan is determined by starting with the insertion areas assigned to each component by the material list analyzer during the initial planning. A three-dimensional (3-D) simulation of this plan is then performed. The initial process plan is modified if problems are discovered in the simulation.

The machine level plan is the sequential order that parts will be inserted on a given machine and the machine configuration that will be used to insert each part. This machine level plan is generated by using the insertion area assigned to each part in the initial process plan and determining all the possible machine configurations. For a given component, the insertion area/machine configuration pair represents one alternative solution in an overall solution space.

Each insertion machine has a device that grips the component and a device that grips the PC board. These will be referred to as the "insertion head" and the "tooling," respectively. The space around the component that is constrained by the insertion head and the space around the PC board that is constrained by the tooling are, in general, functions of the machine configuration. The layout analysis program performs a 3-D simulation of the insertion of each component and insertion area/machine configuration to detect any spatial conflicts between the insertion head, the neighboring components, and/or the tooling.

A mixed graph (Robinson and Foulds, 1980), or "constraint graph," is used to represent the alternative solutions and the constraints between these solutions. Nodes represent alternative solutions, directed arcs represent spatial constraints, and non-directed arcs represent alternative insertion area/machine configurations for a given component. Figure D.4 shows an example of such a constraint graph.

A sequence is generated from the constraint graph by using a cost function that returns the cost to move from any node in the graph to

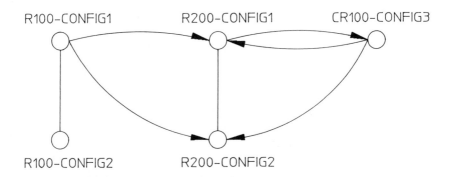

FIGURE D.4 **Example constraint graph**
    In this example, R100 has two alternative machine configurations. R100-config[2] is not constrained by any other parts, but R100-config[1] is constrained by R200. There is a cyclic constraint between R200-config[1] and CR100. Since CR100 has only one alternative machine configuration, R200-config[2] will be chosen.

another. The cost is based on insertion and setup time. A look-ahead technique used in the costing function allows a near-minimum cost sequence to be generated in a single pass.

Certain kinds of spatial conflicts can only be resolved by changing the process plan, for example, cyclic constraints or insertion head spatial conflicts with tooling. In these cases, the process plan is modified by assigning a component to a manual insertion area, where there are minimal spatial constraints.

Rules are used to analyze the final process plan automatically and suggest changes to the layout that could improve the cost and quality. These suggestions provide feedback to the designer in the form of a layout analysis report. Figure D.5 is an example of such a report. The designer may take these suggestions, make changes to the layout, and rerun the layout analysis program. This may be repeated several times until the designer is satisfied that the layout is as manufacturable as possible.

```
LAYOUT ANALYSIS REPORT
RELAXED MODE

BOARD:    01234-66514
REV:      C
DATE:     3/5/1989
-------------------------------------------------------------------

SUMMARY:
 TOTAL PARTS:466
 TOTAL NON-AUTO-INSERTABLE PARTS:131
 NON-AUTO-INSERTABLE PARTS DUE TO LAYOUT:17
 PERCENT AUTO-INSERTABLE: 71.89
-------------------------------------------------------------------

VCD LAYOUT SUGGESTIONS

  (BOARD MUST BE ROTATED TOO MANY TIMES.)
    (MOVE CR209 (1901-0040 D450 X= 10.924 Y= -4.9 (0 0 0 STD-HEAD)))
      (:REASON (TOO CLOSE TO (R244 R245 R250)))

  (POLAR PARTS APPEAR IN TOO MANY ORIENTATIONS.)

  (C048 NOT VCD INSERTABLE)
    (MOVE C048 (0180-0210 C500 X= 3.9 Y= -5.925))
      (:REASON (HOLE SPACING - BODY LENGTH < 0.15 IN.))

  (C037 NOT VCD INSERTABLE)
    (MOVE C037 (0160-0162 A700 X= 2.225 Y= -2.675))
      (:REASON (HOLE SPACING - BODY LENGTH < 0.15 IN.))
```

FIGURE D.5 Example layout analysis report

RESULTS

This automated DFM analysis system has been in use, in various stages of completeness, since January 1988. Improvements in cost, quality, and cycle time have already been realized. The following is a list of some of the results achieved to date:

- The original intent of this automated DFM analysis system was to replace a job that was being done manually by Process Engineers. The goal was simply to do the job at least as well as the human experts, freeing them from this tedious task. The result has been that the system is much faster and does a much more thorough analysis than is possible by its human counterparts. The amount of time spent by Process Engineering has been reduced from approximately one hour per design to ten minutes per design.
- Figure D.6 shows that the average auto-insertion rate, prior to the implementation of this system, was approximately 70%. More re-

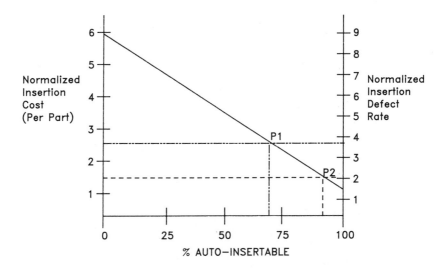

FIGURE D.6 A before-and-after look at auto-insertability
Point P1 represents the current set of PC boards, with auto-insertability averaging approximately 70%. Point P2 represents the set of PC boards that has been analyzed by the DFM system. The auto-insertability of these boards is averaging approximately 90%.

cent PC board designs that have been analyzed by the DFM system are averaging approximately 90%, with some running as high as 95%. This includes both analog and digital designs. If one extrapolates using the graph in Figure D.6, this will result in a 40% reduction in the average cost per insertion. Insertion defects will be reduced by 45%. Since insertion defects are the largest single cause of defects in LSID's PC assembly process, this will result in a marked improvement in overall quality.

In addition to simply moving to a new point on this curve, we also expect that the scale on the "defects" axis will change, causing the 100% auto-insertability point to be much lower. In the past, marginally nonauto-insertable components were often "forced" into auto-insertion, thus pushing the insertion machines beyond their specified limits. This was done in an attempt to auto-insert as many components as possible. The 3-D simulation detects these marginal cases, allowing the designer to eliminate the majority of them.

• One type of defect, not often considered, is the documentation defect (e.g., errors in the material list such as mistyped part numbers or reference designators). Upon implementing this DFM analysis system, we have discovered that nearly every design has some sort of documentation error. Without the system, these errors would have slipped through into manufacturing, causing errors and/or delays in prototypes, and, in some cases, even resulting in a production change order to correct the problem, after release to production.

• The DFM analysis system provides automated enforcement of a preferred parts list. This has proven especially effective for surface mount components. After running seventeen PC boards (mostly analog) containing surface mount components through the system, the total number of unique part numbers was reduced by 6.5%.

• Manufacturing processes are constantly evolving to meet the demands of new product technologies and to improve cost, quality, and cycle time. Because of this, flexibility is a crucial attribute of any DFM system. The use of a model-based approach has proven to be very flexible, allowing changes to be made quickly and easily. Once the original model was created for our radial insertion machine, it took only about one week per machine to develop layout analysis for the other insertion machines in our manufacturing process.

• A side benefit of this DFM system is improved automatic generation of process plans and auto-insertion pattern file programs. Previous to this system, LSID had automatic pattern file generation for most machines, but these programs required a great deal of human interaction. The pattern file programs generated by this system require almost no human interaction. This has resulted in significant productivity improvements. On the average, a savings of 95% of overhead programming time has been reported. For example, programming time for LSID's radial insertion machine has been dramatically reduced from two hours to five minutes. Similar reductions have been reported for the dip, axial, testpin, and surface mount auto-inserters. In addition, production need not be interrupted to perform the program "teaching" that was required on some auto-insertion equipment.

· Automated process planning and pattern file generation have resulted in a new "critical path" for production prototype turnaround. Since the auto- and manual-insertion pattern files are usually completed before the blank PC board is received from fabrication, LSID is now concentrating on reducing the prototype blank PC board turnaround. LSID's goal for turnaround time of lab prototype PC boards, from the time the layout has been completed, is now one day.

· Another unexpected benefit of automated layout analysis and process planning is in the area of process simulation. For example, insertion heads and tooling can be modified and tested using 3-D simulation on a variety of assemblies before any changes are made to the actual insertion hardware. Process plans can be easily altered and tested to determine what effects might result in overall auto-insertability and assembly cost. This type of simulation has been used at LSID to modify a radial insertion head to give greater clearance around neighboring components, thus reducing insertion defects; and in the characterization of design alternatives for proposed PC board process carriers.

SUMMARY

A great deal has been learned from designing and actually implementing this system. Many false starts and pitfalls encountered along the way have not been described in this paper. The following is a summary of the major conclusions that can be drawn from this development effort.

1. Automated DFM analysis is much more thorough and timely than manual DFM analysis.
2. Since a process plan must be generated before suggestions can be made to improve manufacturability, DFM analysis and process planning are inextricably linked. Because of this, automated DFM analysis should be developed as an integral part of an overall CIM system.
3. Flexibility to handle changing and varied processes is a key design criteria for a DFM system. A model-based simulation approach to DFM analysis allows this flexibility.

Future plans for this system include (1) translating the layout analysis program from LISP to C to improve speed and portability, (2) adding a wave soldering analysis module, (3) adding a process carrier analysis module, and (4) developing a similar system for mechanical assemblies. This development is part of an ongoing effort at LSID to gain competitive advantage through improved manufacturability.

REFERENCES

Nicholson, Clark. 1988. *A Knowledge-Based Planning and Manufacturability Analysis System for Printed Circuit Board Designs.* Masters thesis, University of Washington, Seattle.

Robinson, D. F., and L. R. Foulds. 1980. *Digraphs: Theory and Techniques.* New York: Gordon and Breach.

Winston, Patrick H. 1984. *Artificial Intelligence.* Reading, MA: Addison-Wesley.

E

Design for Manufacturing and Solid Modeling*

JOHN M. WALLACH
NCR, Cambridge

ABSTRACT

This paper describes why solid modeling is used for all new product mechanical development at the Cambridge Division of NCR Corporation. It also describes practical methods to promote the implementation of solid modeling in the design process. The use of solid modeling is a key part of the Design for Manufacturing (DFM) strategy used at NCR, Cambridge. Cambridge demonstrated the success that can be accomplished through the combination of solid modeling and DFM strategy on the NCR 2760 Retail Terminal development. The success of that project has led to the utilization of solid modeling for all mechanical development at NCR, Cambridge. This paper describes practical reasons to implement a solid modeling development process utilizing DFM strategy.

*This paper is based on material originally presented at the Second International Conference on Design for Manufacturability, an annual event sponsored by Management Roundtable, Inc., 1050 Commonwealth Avenue, Boston, MA 02215, (617) 232-8080.

INTRODUCTION

NCR, Cambridge, has realized the importance of Design for Manufacturing strategy to new product development. The introduction of this strategy into the development process was accomplished through a corporate sponsored pilot program where "solid modeling" was utilized for the entire Mechanical product definition of the NCR 2760 Retail Terminal.

Early results showed that manufacturing issues were successfully addressed as part of the design process. Cambridge engineers were able to reduce both parts count and assembly time by 80% over the terminal the 2760 replaced. This was accomplished before the 2760 terminal went into production.

Cambridge has implemented a production solid modeling development process based on the DFM strategy. All new Product Mechanical development is accomplished in IDEAS™ (a 3-D design tool). It is appropriate to review the strategy Cambridge used to implement a solid modeling design process, and how this has contributed to the success of DFM.

SOLID MODELING AND DFM

Consider the attributes of the Cambridge development process based on solid modeling:

- All Mechanical detail definition is completed in solid modeling.
- Assembly layouts are completed in System Assembly and transferred using View Dependent IGES (Initial Graphics Exchange Specification, a neutral-format data translation tool).
- DFM team design reviews are conducted at Mechanical Design System workstations.
- Interference checks are completed before design versions are released to vendors.
- GEOMOD™ IGES is utilized for design transfer to vendors, manufacturing, and Technical Publications.

Each of these attributes is important to DFM success. How each attribute fits into the DFM process will be explained in detail.

Detail Definition

Cambridge engineers are convinced that the key to DFM success is to keep the development process (including vendors) centered around the solid model data base. Manufacturing cost reduction, through DFM, helped convince Cambridge engineers to make the transition from a 3-D Wire Frame design methodology to solid modeling. The Cambridge DFM strategy required the entire product design to be completed in the solid model and then reviewed using Boothroyd Dewhurst software. The DFM team must have confidence that the design model they are reviewing will be the one manufactured. Practical experience suggests that the completion of the design in the solid modeler will simplify the DFM review process.

Often, engineers have questioned the effectiveness and ease of change when using a solid modeler. While there is room for improvement in this area, Cambridge engineers utilize program files to quickly implement changes suggested by members of the DFM team. Another issue often mentioned as a deterrent to a production solid modeling design process is the training and learning curve required for successful use. At Cambridge, engineers have been able to make the transition from 3-D wireframe to solid modeling during the implementation of new product designs and still maintain schedules. This has given Cambridge engineers confidence that new technology can be implemented during production schedules. The benefits solid modeling can offer the organization more than offset the difficulties sometimes encountered with its use.

System Assembly

The Cambridge DFM process places equal emphasis on the manufacturing process and the design process. The manufacturing engineer completes the assembly process while the design is still being developed. The Boothroyd Dewhurst software is used to keep score as Manufacturing costs are compared to engineering costs. The System Assembly module in IDEAS allows the manufacturing engineer to visualize the expected assembly process (see Figure E.1).

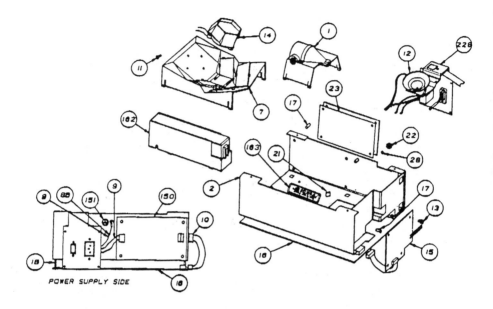

FIGURE E.1 **NCR 7852 scanner assembly**

DFM Design Reviews

Another benefit of IDEAS is the visualization advantage of the solid model. This allows more people in the organization (Safety Engineers, Customer Service, Quality, etc.) to participate in the design process. This is an important example of an organizational advantage of solid modeling that can be used to promote the transition to solid modeling. The poor visualization of traditional 2-D drawings made it difficult to review mechanical design prior to proto-typing. Design changes required after proto-typing were costly and difficult to implement. See Figure E.2. The Cambridge DFM process is based on "real time" design review using the advantages of the IDEAS Solid Model to verify designs before tooling dollars are spent and valuable design time lost.

Interference Checks

IDEAS Interference checks can be a powerful tool to check cabinetry interferences in the assembled position and in the movement neces-

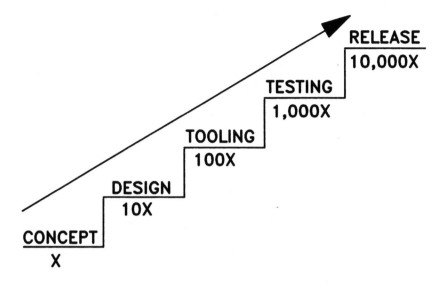

FIGURE E.2 Cost of design changes

sary to assemble. Figure E.3 demonstrates how the System Assembly module can be used to stimulate the service position and the operating position of a scanner product. Interference checks can then be developed for both configurations. If the manufacturing engineer is not confident of the assembly process, any number of configurations or positions can be developed to verify the assembly process. Often, Cambridge designers will do checks as an overnight batch job using program files. The designer can print the results of a system assembly interference check, which then serves as a reference for the DFM team. Interference checking has given designers confidence that they can successfully design plastic parts without the aid of soft tooling. This also eliminates the development time needed to make the transition from soft tools to production mold tools.

Electronic Transfer and Documentation

The use of IDEAS for design and the transfer of IGES IDEAS data to vendors changes documentation requirements. Designers need only create application specific (inspection, tech pubs, assembly) drawings. See Figure E.4. Because mechanical part design is transferred

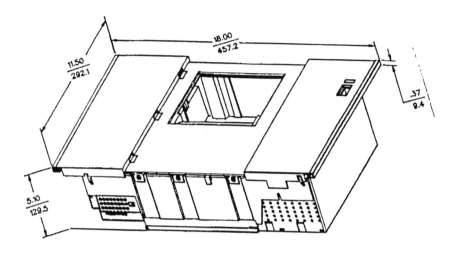

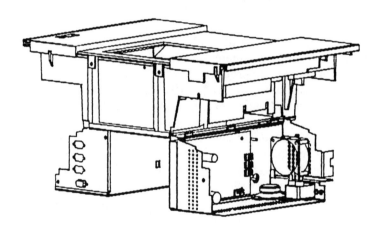

FIGURE E.3 NCR 7851 scanner

to vendors through IGES, on electronic media, engineering can avoid committing development resources to complete detailed engineering drawings. Instead, engineering can commit development resources to optimizing the statistical process control data (control dimensions) to be used by the vendor to control quality.

The Technical Publications Department uses IDEAS and IGES

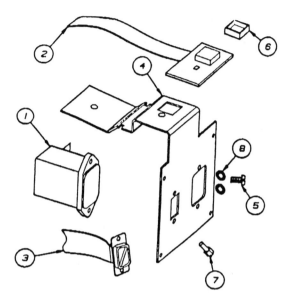

FIGURE E.4 NCR 7852 scanner I/O plate

for the development of illustrations and parts manuals. It has been estimated that this has cut Technical Publications parts manual development time in half. This process has been successful to the point where Technical Publication personnel now use IDEAS to develop assembly configurations or alter configurations developed by designers (Figures E.1 and E.4 are examples of this).

Product Management is also starting to realize the advantage of using IDEAS graphics in their marketing publications. This is another example of how IDEAS creates organizational advantages. IDEAS has allowed Cambridge to simplify the documentation process and provide productivity improvements for the organization.

INTANGIBLE BENEFITS OF SOLID MODELING

The Cambridge organization has gained many benefits because engineering develops complete solid models for all product mechanical designs. When first introduced to solid modeling, engineers at Cambridge often wondered if the additional time needed to develop complete solid models was justifiable. Engineers are under constant pressure to shorten the development cycle while still maintaining excellent quality. This paper has presented tangible benefits that can be gained through solid modeling. The manufacturing improvements for the two scanners shown in this paper are as follows:

Manufacturing Results for NCR 7852 Scanner
(versus NCR 7824 Scanner/Scale)

   68% reduction in part count

   75% reduction in assembly time

   70% reduction in screws

Manufacturing Improvements for NCR 7851 Scanner
(versus Sheet Metal Release)

   60% reduction in parts

   70% reduction in assembly time

   69% reduction in fasteners

These are benefits that will actually shorten the time from product concept to manufactured product and reduce manufacturing costs. There are additional benefits for the engineering manufacturing organization from the use of solid modeling for mechanical design.

Cambridge engineers have always known that significant advantages could be gained if the "wall" between engineering and manufacturing was removed. In the past, it was difficult to promote cooperation between engineering and manufacturing. Engineers concentrated on functionality and quality, occasionally consulting manufacturing, and then handed finished designs to manufacturing. Manufacturing engineers took pride in being able to put into production whatever engineering passed over the wall.

Solid modeling under the Design for Manufacturing strategy has changed this process. Manufacturing engineers now have the opportunity to optimize the manufacturing process as part of the DFM team during the design cycle. There is a feeling of ownership for the product design by the entire DFM team, which includes individuals from every part of the organization. The success of the product is shared by the team. The availability of the design for review, through the solid model, is a major contributor to the "cultural" change necessary for successful development by a DFM team.

SUMMARY

NCR, Cambridge, has placed increased emphasis on teamwork and manufacturing. Engineers from different disciplines and backgrounds work together to improve quality and reduce the time to get the product to market. At Cambridge, products are developed using the Design for Manufacturing strategy in order to meet the challenge of lowering costs and improving quality. The objective of the Cambridge DFM strategy is to design manufacturability into the product before the design is proto-typed or tooled. This strategy requires parallel development, which emphasizes cooperation.

Product engineers work within the DFM team concept by utilizing the benefits of IDEAS and Boothroyd Dewhurst software. The solid model is a very visual, accurate representation of potential design. The entire product design is completed in the Solid Modeler, and this data base is used for fabrication processes. Through the utilization of solid modeling under the DFM strategy, the entire organization benefits from the higher product quality resulting from the process.

F

Concurrent Product/ Process Development: The Role of Suppliers*

Robert A. Carringer

International TechneGroup, Incorporated

In today's global market, competition for market share may come from across town or across the ocean. World-class quality and cost requirements—coupled with the critical need to shorten product development time—force companies to seek new business relationships with their suppliers to remain competitive. These new relationships often involve distributed engineering concepts, the concurrent development of products, and their associated manufacturing processes.

*This paper is based on material originally presented at the Second International Conference on Design for Manufacturability, an annual event sponsored by Management Roundtable, Inc., 1050 Commonwealth Avenue, Boston, MA 02215, (617) 232-8080.

259

IMPROVED COMMUNICATIONS SHORTEN
PRODUCT DEVELOPMENT TIME

Manufacturing companies increasingly view suppliers as critical resources that must be carefully managed, as well as evaluated, in terms of their strategic contribution to a company's market position. They are reducing the number of critical suppliers to improve efficiency in management, engineering, and purchasing. Manufacturers are asking their suppliers to participate earlier in the product development process to reduce lead time, improve quality, and reduce cost. The new trading partner relationships which result are being facilitated by the introduction of advanced information technology and high-speed communication links.

INTRODUCTION TO CP/PD

Concurrent Product/Process Development (CP/PD) is a seven-phase methodology used to shorten the product development time, improve the quality designed into the product, and reduce overall development and product costs. The product alternative concepts are developed following a rigorous examination of customer requirements. Each product alternative is evaluated from a manufacturing process perspective. Both product and process alternatives are evaluated through simulation, cost modeling, and testing prior to detailed product/process design. For this process to be effective, suppliers providing sub-contracted subsystems and parts must also participate in the concurrent product and process design. Suppliers need to be selected by the prime manufacturing company earlier in the product development process. The relationship between the supplier and the manufacturing company takes on fundamentally new dimensions.

This concept paper describes the role of suppliers in CP/PD and how it is different from typical relationships suppliers have with manufacturing companies.

TRADITIONAL ROLE OF SUPPLIERS
IN PRODUCT DEVELOPMENT

Suppliers typically enter the product development process after the product design has been conceptually developed, analyzed and tested, and detailed. The suppliers receive part prints and specifications from the manufacturer, who is their customer. Suppliers work with the manufacturer's Purchasing Department to deliver a quote. The manufacturer selects the supplier from a couple of sources, all of which have gone through the same quoting procedure. Following selection and initial production, the supplier usually contributes its expertise along with that of Purchasing to reduce the cost of the parts. This cost reduction occurs through suggested engineering changes to material, processing or, maybe, geometry (part design). The supplier or Purchasing rarely makes drastic changes to the fundamental design concept.

ROLE OF SUPPLIERS IN CP/PD

Suppliers are selected earlier in the product development process in Concurrent Product/Process Development. Strategic suppliers, those who provide major subsystems, are selected during Phase II, Alternative System Concepts, or at the beginning of Phase III, Alternative Subsystem Concepts.

Strategic suppliers are selected by the manufacturer on the basis of design/manufacturing technology, willingness to participate in the CP/PD procedure, and management commitment to the prices allocated to the subsystem from the market modeling. Essentially, the manufacturer must ask:

- Can we find a supplier using more advanced technologies for design or manufacturing anywhere in the world?
- Can we find a supplier who can deliver consistently the same or better quality product at a lower price?

If the answer to both of these questions is "No," then the manufacturing company selects the strategic supplier of subsystems.

Strategic suppliers participate in the CP/PD design process beginning in Phase III, Alternative Subsystem Concepts. The supplier uses its engineering talent to sit on the product development team at the manufacturer. The cost reduction activity normally found in the traditional supplier role becomes *cost avoidance* through better designs—*getting it right the first time*.

To assist the supplier, technologies for product definition data exchange and data communications combine to form a link between the supplier and manufacturer. The purpose is to improve communications between the two trading partners, thereby reducing the time consumed through normal methods.

CUSTOMER/SUPPLIER HUB FOR CAD/CAM/CAE DATA TRANSFER

The objective of the Customer/Supplier Hub is to replace the paper transfer of engineering and manufacturing information between the supplier and the manufacturer. This information includes engineering drawings or models, specifications, and other information in alphanumeric or graphic formats. The Customer/Supplier Hub has portions of the components shown in Figure F.1.

CAD/CAM/CAE data exchange is validated between the manufacturer and every supplier possessing a unique CAD/CAM/CAE system. Typically, IGES is used for exchange between dissimilar systems. The IGES interchange must be tested for accuracy and loss of data. Some lost data may not impact the information exchange process (such as dimensions) while the loss of other types of data (such as geometry) may result in reduced productivity.

Some companies consider moving all IGES translation to one

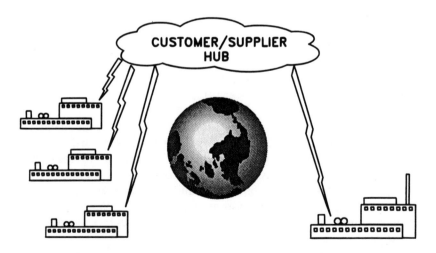

FIGURE F.1 Customer/supplier hub

centralized location on the basis of equipment utilization, dedicated IGES personnel, and the logistics of supplier communications. Central site IGES processing can also provide a service to company users to check the accuracy of the translation. The IGES expertise can be made available to suppliers who also will experience problems with using IGES.

A transmission network for data communication is established through the manufacturer's internal network or through a third-party, public network like the GE Information Services Design*Express Network. Using a network alleviates the burden on the manufacturer to support the end-to-end communications that otherwise need to be in place. Networks provide international access, security, and high-speed communication alternatives.

For suppliers not currently using CAD/CAM/CAE, a supplier workstation needs to be provided. The Distributed Engineering Workstation (DEW) is a combination of a PC-based CAD/CAM system, IGES translation and flavoring software, and communications software. These packages are interfaced and checked out so the supplier can rely on its effectiveness. The DEW is a communications gateway primarily for the supplier. However, depending on the supplier's needs, alternative software needs to be provided on the system. For example, to participate in CP/PD, the supplier must have access to project management software, QFD requirements capture software (such as QFD/CAPTURE™), and technical publishing of software.

Finally, for the manufacturer and suppliers, training of several types is needed to ensure productivity from the beginning of the project. Supplier personnel need to understand their impact on the CP/PD process through knowing the process and their role within it. The methods for communication also are an element of the training required. Support for the IGES translation and the data communication is necessary for the smaller suppliers as well as some of the functional organizations within the manufacturer.

In operation, the actual needs for information interchange became evident through an analysis of the network traffic. For the initial six-week period of a supplier communication activity, the information exchanged included process specifications, manufacturing equipment specifications, databases containing administrative information about tests and analyses and letters. In this case, the amount of non-CAD/CAM data was underestimated.

G

Quality Function Deployment: The Latent Potential of Phases III and IV

L. P. SULLIVAN
American Supplier Institute

INTRODUCTION

Quality Function Deployment (QFD) is expanding very rapidly in U.S. industries. While many individuals and companies have contributed to this movement, we owe special thanks to Don Clausing (MIT) who introduced Quality Function Deployment to Ford Motor Company in June, 1984. Also, we are thankful to the Budd Company and Kelsey-Hayes, which prepared the first practical applications of QFD for U.S. companies. ITT is also recognized for its contribution in the development of training material to deploy QFD methodology into the production workforce.

What is the value of QFD to U.S. companies, to consumers, and to the nation as a whole? How much of this value has been realized and what is the future potential? The most important questions relate to the latent potential of Phases III and IV, which must be implemented before the full value of QFD can be realized. In this paper, we will assess the value of QFD and the latent potential yet to be realized.

VALUE OF QUALITY FUNCTION DEPLOYMENT

Although most Japanese companies use some form of QFD with significant benefits, U.S. companies stand to gain a greater impact from this technology. The value of QFD is to deploy the voice of the customer horizontally from product development through manufacturing quality control. This is accomplished through four phases of QFD, which are illustrated in Figure G.1 and outlined in the following.

PHASE I: PRODUCT PLANNING

The initial phase of QFD is very popular in the United States, and most of our current activities center around the basic "House of Quality." While Phase I is the most important phase from the standpoint of defining customer wants in relation to product parameters, its power can only be realized through the deployment of these design parameters through Phases II, III, and IV.

PHASE II

This phase is associated primarily with product engineering functions, where design parameters are transferred into part characteristics, and, more importantly, the creation of target values, which represent the best values for fit, function, and appearance. At this point in time, I would estimate that approximately half of the applications in U.S. companies have progressed through Phase II.

PHASE III: PROCESS PLANNING

Most of the QFD activities covered in Phases I and II are performed by product development and engineering activities. When we enter

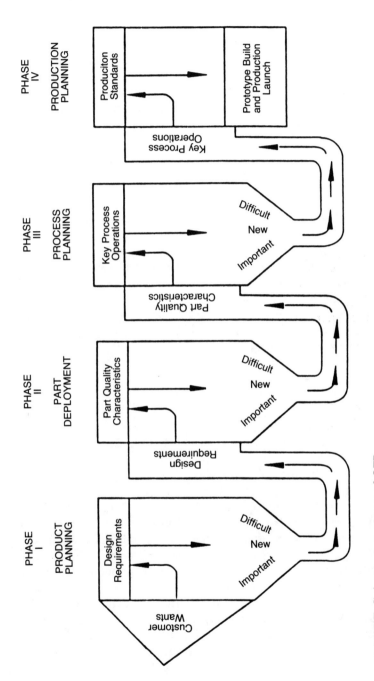

FIGURE G.1 The four phases of QFD

267

Phases III and IV, it is necessary and even essential to involve floor-level process engineers, production supervisors, and line operators. Phase III is where target values from the product engineering phase have been deployed into process parameters for manufacturing and assembly. It is also in this phase where process capability levels are developed and activities established for continuous improvement. Today in the United States, very few QFD applications involve process planning.

PHASE IV: PRODUCTION PLANNING

This last phase in QFD transfers target values from process planning into production standards, which are necessary to maintain and integrate production and assembly activities. This is the final stage in the linkage between the voice of the customer in Phase I through subsequent stages in engineering, manufacturing, and on-line quality control. With this last stage, all employees of the company and their daily functions are firmly linked to the customer, and all activities interact to achieve customer expectations (i.e., cost, quality, comfort, style, performance, and delivery).

QUESTION

Why is it that U.S. companies stand to benefit more than Japanese companies through the application of QFD? In Japan, managers, engineers, and workers are more naturally cross-functional, cultural aspects tend to promote group effort, and concensus thinking. In the United States, we are more vertically oriented with strong management motivated to sub-optimize for individual and/or departmental achievements. The U.S. business culture tends to promote breakthrough achievements which, in many cases, inhibits cross-functional interaction. The QFD advantage to U.S. industries is to maintain the breakthrough culture with emphasis on continuous improvement through more effective cross-functional interactions. If we can do both, then U.S. companies will realize an important competitive advantage over Japanese companies. Therefore, QFD offers a far greater potential for U.S. companies than for Japanese companies. However, a serious problem remains for U.S. management to fully understand the latent potential of Phases III and IV and to deal with barriers which prevent the full benefits of QFD.

LATENT POTENTIAL

While the benefits being realized today are important, they are based primarily on Phases I and II (product planning and part deployment activities) since, for the most part, only engineers and managers have been exposed to QFD. Future benefits are far more significant in Phases III and IV (process planning and production planning) due to the sheer strength of manpower and knowledge which is available.

Of the several thousand QFD case studies underway today in U.S. companies, the majority have not progressed fully through Phases III and IV. We seem to have reached an impasse due to the difficulty in transferring target values from design to production. In some cases, the division between management and labor has contributed to this impasse. U.S. industrial labor contracts, and the tenderness of issues between management and labor are contributing to and, in some cases, preventing deployment through Phases III and IV. This is a very complex issue and well beyond my experience and knowledge. However, I do understand the power of this latent potential where the majority of human resources are under-utilized.

In the typical U.S. industrial company, 5% of the employees are engineers, 5% are managers and executives, another 10% are in administrative functions, and the balance (80%) are floor supervisors and workers. By the nature of QFD, Phases I and II involve engineers and managers, while Phases III and IV involve supervisors and floor workers. Therefore, 80% of the most knowledgeable human resources are not being utilized to their full potential. By this, I mean their mental capacity has not been integrated to fully realize the voice of the customer. Figure G.2 illustrates the significance of this latent potential. Why is it that U.S. management puts so little value on the mental capacity of floor workers? This question has always bothered me, and I do not have a good answer.

Recently I received a copy of a speech to U.S. executives by Mr. Konosuke Matsushita in early 1988. It is entitled, "We Win—You Lose." The following comments by Mr. Matsushita may provide some insight into the dilemma facing U.S. industry.

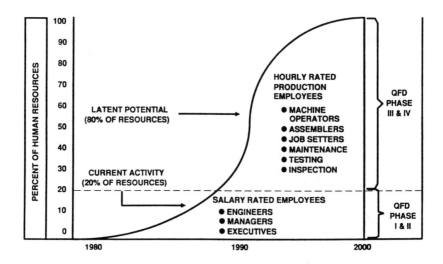

FIGURE G.2 Human resource allocation of QFD

We will win, and you will lose. You cannot do anything about it because your failure is an internal disease. Your companies are based on Taylor's principles. Worse, your heads are taylorized too. You firmly believe that good management means executives on one side, and workers on the other; on one side, men who think, and on the other side, men who can only work. For you, management is the art of smoothly transferring the executives' ideas to the workers' hands.

We have passed the Taylor stage. We are aware that business has become terribly complex. Survival is very uncertain in an environment increasingly filled with risk, the unexpected, and competition. Therefore, a company must have the constant commitment of all its employees to survive. For us, management is the entire workforce's intellectual commitment at the service of the company . . . without self-imposed functional or class barriers.

We have measured—better than you—the new technological and economic challenges. We know that the intelligence of a few technocrats—even very bright ones—has become totally inadequate to face these challenges. Only the intellects of all employees can permit a company to live with the ups and downs and requirements of its new environment. Yes, we will win and you will lose. For you are not able to rid your minds of the obsolete Taylorisms that we never had.

I am not so sure I agree with Mr. Matsushita because I think the division between management and labor can be resolved, and I think QFD is the mechanism to improve cross-functional interaction. Our efforts in the coming years, therefore, should be concentrated on Phases III and IV to complete the linkage between product planning and production planning. This can only be accomplished through cross-functional horizontal training and project applications involving production workers. After many years of working on quality method applications, we now realize that training can be a pitfall to quality improvement if it is not accompanied by project applications. In fact, we feel most of the learning comes from applications and very little from lecture-type training. Also, the use of outside consultants can be a pitfall as well. Although some awareness training may be required from outside consultants, it is far more effective for project applications to be facilitated by internal employees as a part of their regular jobs. In QFD Phases III and IV, it is essential that these applications be facilitated by labor (i.e., hourly workers) and not by management.

In the typical U.S. company, training for engineers and training for workers carry a significantly different material content. Shown in Figure G.3, for example, is a large automotive company comparison between training for engineers and training for production workers.

You can see where very little cross-linkage exists and, therefore, a common base of knowledge is lacking. The important objective of training is to provide a common base of knowledge to facilitate technical interaction between engineers and workers. This technical interaction tends to neutralize emotion and promote more concrete efforts for continuous improvement. Several U.S. companies have been very successful at promoting cross-functional interaction through education and training. The Allison Transmission Division of General Motors is an outstanding example where UAW employees develop the training material, conduct classes, and facilitate project implementation.

Another good example is the Transmission and Chassis Division of Ford Motor Company. The Ford program is oriented around project applications with a well structured Dimensional Control Program (DCP). The Ford Transmission and Chassis DCP program was not imported from Japan, but developed by UAW hourly workers in cooperation with engineers and management. It focuses on process capability improvement as related to QFD target values deployed through Phases III and IV. This is one of the very few production programs in U.S. industries that provides a direct linkage with the voice of the

ENGINEERS	PRODUCTION WORKER

ENGINEERS	PRODUCTION WORKER
1. OVERVIEW OF VARIABILITY	1. THE PRODUCT LINE
2. INTRODUCTION TO QUALITY METHODS	2. THE CALCULATOR
3. 5 PHASE PROBLEM RESOLUTION PROCESS	3. MATH ENHANCEMENTS
	4. MICROMETERS AND CALIPERS
4. SPC CONTROL CHART OVERVIEW	5. GEOMETRIC DIMENSIONING AND TOLERANCES
5. SPC OVERVIEW	
6. SPC TRAINING	6. ROBOTICS
7. INTRODUCTION TO PROBABILITY AND STATISTICS	7. THE ORGANIZATION STATE OF BUSINESS
8. WEIBULL ANALYSIS	8. STATISTICAL PROCESS CONTROL
9. ACCELERATED EXPERIMENTS	9. QUALITY ASSURANCE
10. RELIABILITY GROWTH	10. FINANCIAL OVERVIEW
11. TAGUCHI OVERVIEW	11. THE MACHINING PROCESSES
12. PARAMETER DESIGN TECHNICAL OVERVIEW	12. ASSERTIVE COMMUNICATIONS
13. PARAMETER DESIGN TEAM TRAINING	13. HYDRAULICS ELECTRICAL PNEUMATICS
14. DOE RESIDENT CONSULTANT AND TRAINER TRAINING	14. BLUEPRINTS
15. QUALITY FUNCTION DEPLOYMENT OVERVIEW	15. MEASUREMENTS LAB
	16. THE MACHINING PROCESSES
16. QFD WORKSHOP PART I PRODUCT PLANNING	17. GREEN MACHINE PROGRAM
17. QFD WORKSHOP PART II PART DEVELOPMENT	18. EVALUATION
18. QFD WORKSHOP PARTS III & IV PROCESS AND PRODUCTION PLANNING	

FIGURE G.3 **Training for engineers and production workers in a large automotive company**

customer, and involves both engineers and production workers. The Ford Transmission and Chassis DCP project could very well form the basis for all Ford suppliers to improve process capability around specific engineering parameters and target values.

An interesting point we learned from Nippondenso in Japan is the importance of personal capability. This is the appraisal of production workers and their personal capability relative to quality method applications. Figure G.4 is an appraisal sheet for personal capability. Factor item 6 (data related), for example, shows knowledge of process capability and its improvement as an important criteria for personal capability.

When looking at results and targets for this employee, we see where personal improvement is required. In these cases, the foreman and section head are required to develop and facilitate an annual improvement program for these individuals. Management thinking at Nippondenso focuses on personal capability as a priority project for management, and this relates to the most significant weakness in U.S. companies. By weakness, I mean the lack of training and project implementation for production workers to integrate their activities with engineering, and realize the full benefit of QFD Phases III and IV.

CONVERTING QUALITY COST

One of the main inhibitors to the full extension of QFD in U.S. companies is the great difficulty (on the part of management) to understand the huge potential for cost savings through quality improvement.

In March 1982, during my first Study Mission to Japan, I learned about the Taguchi Quality Loss Function while reviewing case studies at Nippondenso. Upon my return, I made a video tape using the cigar lighter example from Tokai Rika, which became quite popular in discussing variability as related to control charts. I also conducted many reviews with top management on the Quality Loss Function to illustrate its power in driving quality improvement.

Here we are seven years later, nine additional study missions to evaluate Japanese applications with countless presentations, and today management is only mildly curious about the Quality Loss Function. Why this is, I do not know. It must be associated with the dominance of financial oriented executives who were trained in traditional U.S. cost control systems and, therefore, cannot understand how costs can be reduced by reducing variability. Furthermore, even if management *can* make the transitions (mentally), the corporate financial systems are so ingrained that few executives are willing to mount the challenge necessary to recognize Loss Function calculations in decision making. It is my feeling that production workers with

For Qualification Education	Apraisal of Personal Capability for S A Process		7·16·76	Sect. Hd.	Foreman (appr)			Sect. Hd.	Foreman (appr)	
			Qualified For Ⓢ Proc.				Qualified For Ⓐ Proc.			

Dept.	Manufacturing	Job No.		Name		Asst. to Foreman	Qualified Date	

	Factor Item	Requirements (guidelines)	Unsat. 1	Good 2 3 4	Excel. 5

Skill Standards

1. Operational skill
   1) Sure & fast operation of apparatus & speed change levers of machine tools
   2) Sensitive & fast reaction to abnormal operation of machine tools counterplan

2. Skill for job planning
   1) Fast & sure changing apparatus, dies based on guidelines & knowledge of preparatory arrangement
   2) Sure processing within set time
   3) Sure seizure of points in process

3. Skill in cutting (grading)
   1) Fast & sure setting apparatus & workpieces
   2) Cutting (grinding) under ideal condition described in guidelines
   3) Prediction of abnormal machine tools by judging abrasions & lifetime appropriate countermeasures
   4) Skill to develop process capability of machines to their full capacity. In case of abnormality, identify causes & prevent recurrence

4. Skill in measuring
   1) Proper seleciton of measuring instruments Correct & fast measurement
   2) Sense of judgement of roughness Looseness based on JIS

Standards for QA Capability

5. Standard related
   1) Knowledge & following of job guidelines
   2) Counterplan, proposal for revision of imperfect job guidelines
   3) Knowledge of QC Process Chart, its application to actual situation to find problems

6. Data related
   1) Knowledge of calculation of process capability index & its improvement
   2) Factor analysis of characteristics
   3) Proper judgement of process stability based on control chart
   4) Knowledge of names & their application of basic QC methods

Safety

7. Safety
   1) Alert to safety

Other Majors

8. Knowledge of control guidelines

KS002-0002 1) parts, processes by importance	KS041-0006 indication of division of parts, processes by imp.	KS943-0001 general guidelines for control of major safety parts

Results ⟶   Targets X—X

Process control guidelines for Ⓢ Ⓐ parts (Man. Dept.) Established 5.16.76 (excerpt)
1. Jobs on Ⓢ process are performed by workers with 1st or 2nd grade technician status
2. Technicians for Ⓐ process are approved by supervisor section head

Evaluation date	7.20.76	Score	76

Field for Training	Achievement (expected level)	Training Method	Training Period	Trainer	Texts, training materials	Actual Achievement	Remarks
Fields where training is indespensable to attain an expected level	Level which trainee is expected to attain	Methods For Training	Target Date	Who Trains	What kinds of teaching materials	What is achieved out of set targets for training	Re-training necessary & fields methods
(6) Date related fields	(4) Level	Man to Man	( ) Days ( ) Wks 1 Mo. 2 Mos. ( ) Mos.	Foreman Unit Hd Group Leader QC Pers. Members Staff	Job Guidelines QC Proc. Chart Technical Standard Provision Cards (QC Methods) ( ) ( ) ( )		

FIGURE G.4 Appraisal sheet for personal capability

the support of labor unions can impact change in the cost control system where mid-level managers have failed.

LEARNING THE QUALITY LOSS FUNCTION

Last week I watched Lee Iacocca on TV speaking in Washington before the National Association of Manufacturers. His main message was the poor education system in U.S. schools. One of Mr. Iacocca's comments related to Chrysler corporate training material, which "is written at the sixth grade level because that's all they could handle." He further stated that "If you don't have people who are smarter than the robots they work with, the game is over."

I think Mr. Iacocca should make a distinction between educational levels and employees ability to contribute. Although I agree that more education is always better, I think we should start today and take advantage of existing worker intellect through some creative management. In other words, we must figure out how to "stay in business during alterations" because it may be another twenty years before public education improves.

The issue is how can we create an environment and adopt methods that can be more successful given current educational levels. Engineers and workers alike need simple methods and leadership to facilitate applications. Although the Quality Loss Function (Figure G.5) is based on eighth-grade math, all of the calculations can be performed by hand-held computers. Production operators need only know the target value (voice of the customer), process dispersion from target, and the dollar loss provided by management. Education required to use the Loss Function already exists in the production workforce, but management needs to facilitate its use!

Let's not blame the education system or the worker; today's production worker is more capable than some engineers in understanding target values and improving quality. If management can provide workers with the environment to use their knowledge, we would see real progress in the evolution of QFD in Phases III and IV. If workers could share the financial reward of improving quality (e.g., Loss Function bonus), then we would see hundreds of case studies (applications) with benefit to the customer, to the company, and to the production worker.

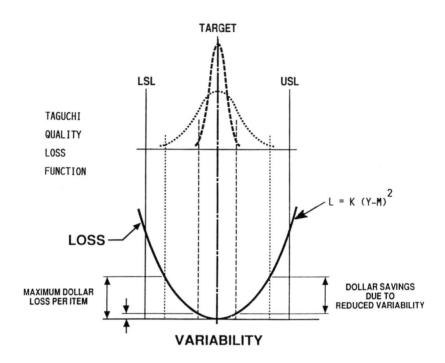

FIGURE G.5 Quality loss function

EXAMPLES

Shown in Figure G.6 are data from Toyota Autobody, the Japanese Company which pioneered QFD in the automotive industry. In 1975, Toyota Autobody was producing 430,000 units per year of a given type with an in-plant repair/finesse cost of $71 per unit. This equated to an annual cost of $30,530,000, or 1.5 times their annual profit. The objective, therefore, was to increase profit by reducing in-plant repair/finesse operations—a fairly simple transformation in management thinking. However, it was essential to use the Loss Function to convert quality improvement to cost savings. Also, it was essential to utilize plant workers to develop the ideas and participate in QFD Phases III and IV.

Shown in Figure G.7 is a similar success story from Toyota Autobody activities in QFD. From January 1977 through April 1984,

Repair Cost Per Vehicle

$71

Annual Vehicle Production

430,000 Units

Annual Repair Cost:
$71 x 430,000 = $30,530,000
1975 Profit = $20,000,000

Repair Cost = 1.5 x Profit

FIGURE G.6 1975 Production figures from Toyota autobody

it realized a 61% reduction in production launch cost by involving floor workers in Phases III and IV.

CONCLUSION

In closing, I would like to further reiterate the importance of Phases III and IV and comment on the latent potential through worker involvement. In doing this, I will relate a quote attributed to Robert Reich. "Workers at all levels add value . . . not by tending machines or carrying out routines, but by continuously discovering opportunities for improvement in products and processes." Also, I will re-state the primary management task, which is to improve the capability of individual employees. This includes creating a common base of knowledge for cross-functional interaction and also the system of linkage to deploy target values through QFD. The essence of our activities to promote continuous improvement, and the latent potential to realize this objective can only flow through a full and complete extension of QFD from Phase I through Phase IV.

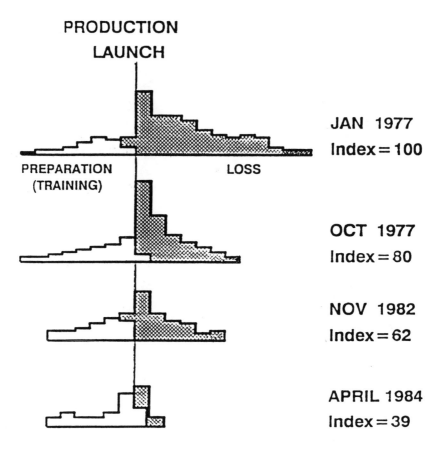

FIGURE G.7 Toyota production launch costs from January 1977 to April 1984

The most important aspect, however, is in the conversion of quality improvement to cost reduction. Without this conversion, future progress and more importantly, worker participation will not happen. The only mechanism for conversion is the Taguchi Quality Loss Function, and U.S. corporate executives (in finance as well as engineering and manufacturing) must understand its power!

It became equally important for workers and union management to understand the Loss Function. Profit sharing is a great motivator for U.S. workers, as well as executives, and programs should be established to enable workers to share in cost savings through quality improvement. I recommend that cross-functional teams of salaried and hourly workers be financially rewarded for their quality achievements. Existing profit-sharing agreements (e.g., U.S. automotive industry) should be amended to include incremental bonus awards for quality improvement using the Quality Loss Function to quantify corporate savings.

Only through mutual financial benefit can salary and hourly workers be integrated. The extension of QFD Phases III and IV into the workforce can be the mechanism to fully realize customer expectations in the product and employee aspirations in the work place.

H

The Implementation of a Participative Management System at the Tektronix Circuit Board Plant

TEKTRONIX CIRCUIT BOARD DIVISION

INTRODUCTION

Whether called employee involvement or labor–management "jointness" or, as it will be referred to here, "participative management," there is a movement spreading throughout U.S. manufacturing companies that is changing the way management and workers approach and perform work. Its implications are broad and significant. Participative management represents one of U.S. industries best hopes for global competitivenes.

Participative management is driven by the fundamental belief that an informed workforce committed to producing superior results and meeting customer needs will be more productive and produce higher quality products. It requires workers to change their old ways of working and gives them greater autonomy and a voice in the decision-making process. It reduces managers' control in the traditional sense, but gives them a powerful tool for improving business performance.

One of the most advanced examples of a participative management system is the Tektronix Circuit Board manufacturing plant, located in Forest Grove, Oregon. In 1983, under the direction of a new manager deeply committed to the principles of participative management, the plant began shedding its old way of doing work—characterized by hierarchical management and assembly-line type production—and undertook the often painful process of change.

The results have been worth it. Increased productivity and quality improvements have driven the plant's sales from $27 million in 1983 to $57 million in 1989. A captive vendor in 1983, now 50% of its sales are from non-Tektronix customers. The plant has almost tripled in size, from less than 300 to 800 employees today, at a time when the company's overall employment levels have dropped from 24,028 (in 1981) to 16,085 (in 1988).

This paper will discuss the process of implementing participative management practices in the plant's journey toward cultural transformation.

BACKGROUND

Founded in 1946, Tektronix, Inc., was founded by two Oregonians to build an oscilloscope that would be faster and more advanced than any currently available. From that beginning, the company has grown to a $14 billion manufacturer of electronic products and systems in the areas of test and measurement, computer graphics, and communications.

The base component driving the company's sophisticated test and measurement tools and information display products is the circuit board.

Until 1983, Tektronix circuit boards were manufactured at three sites on the company's Beaverton, Oregon, campus. In the mid-70's, Tektronix was in the midst of an unprecedented growth spurt that would see sales increase from less than $170 million in 1970 to almost $900 million by 1979, and the number of employees more than double during that period. By 1977, it became apparent that the circuit board facilities could no longer keep up with customer demand. The result was delayed deliveries of products to customers, which, in turn, hurt sales.

In 1979, the decision was made to build a new facility, where all circuit boards would be manufactured, at an estimated cost of $32 million. After the project gained final corporate approval in 1981, the planned startup date was set for January 1983.

At the same time, managers involved with the design of the new plant intended to implement a different set of management practices as well. The result, they hoped, would be a facility that achieved better business performance—including on-time delivery and higher quality and productivity—by more effectively developing and utilizing the skills of its people.

There were many forces in the early 1980's driving managers at Tektronix, as well as at other high-technology companies, to re-evaluate their highly centralized and bureaucratic management structure. These forces included: the extremely competitive nature of the company's markets; the increasing rate of technological change; the changing expectations of a more highly educated workforce; increased job alienation; the under-utilization of employees' creative and intellectual abilities; and the organization's need for increased adaptability and flexibility (Belgard et al., 1988).

Six key elements characterize the management system put into practice in the new plant at Forest Grove. As is discussed in the following, participative management has played a critical role in each of them.

THE FOREST GROVE MANAGEMENT SYSTEM

1. The full plant concept
2. Commonly held values and business understanding
3. Individual and group development
4. Special vendor relations
5. Emphasis on innovative behavior
6. Leadership and support

Full Plant Concept

The "full plant concept" was a critical aspect of Forest Grove's evolution as an environment that fostered the behaviors that characterize participative management. At the old circuit board facility, business functions such as finance, research and development, human resources, and marketing were centralized with those of the company's other component operations. The result was a bureaucratic system that stifled the experimentation and innovation critical to success in the fast-changing world of the electronics industry.

One response to the need for change was a corporate-wide drive toward divisionalization in the early 1980's. The impact of this process for the components facility, including the circuit board operation, was the company goal that they change from captive vendors (producing circuit boards only for Tektronix) to stand-alone businesses.

Management recognized that the ability to respond to the unique needs of its customers was critical to ensuring the plant's success as an independent business. This autonomy enabled the plant to implement and evolve its participative management system free of the way things were done in the rest of the company. The fact that the new plant was situated several miles from the company's Beaverton campus was also symbolic of its new independence.

When the new plant opened on schedule in January 1983, it was already a far leaner operation than the old one, having reduced the number of employees from about 670 to 275. A stagnant growth rate for Tektronix in the early 1980's meant that the new plant had more capacity than the company could utilize. Cultivating non-Tek customers would be a considerable challenge; the old circuit board operation did not have a reputation for high-quality products or on-time deliveries. In fact, several Tek divisions were turning to other circuit board suppliers.

Commonly Held Values and Business Understanding

An integral factor in the plant's metamorphosis was the hiring of a plant manager, Gene Hendrickson, who, from the beginning, had a very clear vision of what the plant could become and saw participative management as the tool for achieving it. From the very beginning, the Forest Grove plant would incorporate his business excellence beliefs. His mission for the business has not changed since it was written in 1982.

Mission Statement

1. To develop the "full plant concept" at Forest Grove.
2. To establish and maintain the position of world leadership in the design and manufacture of high-technology interconnect systems.

3. To assure Tektronix holds the competitive advantage worldwide in circuit board technology, availability, cost, quality, and time to market.
4. To make a significant contribution to the financial success of Tektronix.
5. To achieve this mission while maintaining a high quality of working life.

To significantly improve productivity and achieve its business goals, Forest Grove management believed it was imperative that every employee:

- Understands the customer's need
- Has knowledge of the customer and his business
- Has information about competitors
- Has a basic understanding of the business
- Has access to operational information
- Can influence change

Both formal and informal channels are utilized to communicate this information and values to employees.

Perhaps the most significant formal communications channel is the Core Group. The Core Group was originally formed as a problem-solving group whose objective was to get the new plant up and running. Today its major function is to provide to the entire plant on a daily basis, comprehensive, up-to-date operational information and to provide a forum for addressing issues that require immediate attention. One major benefit of Core Group meetings is that they enable any manager who has been away from the plant to be quickly apprised of the status of the various operations of the business.

The Core Group is comprised of technicians, managers, and support personnel, and meets daily, usually for less than one-half hour, ninety minutes after the start of each of the three shifts. The information shared at the meeting is then disseminated by each representative to their group.

In addition to sharing information and providing better understanding of other disciplines and groups in the plant, Core Group meetings give their participants the opportunity to hone their presentation and speaking skills. Each Business Element and support group determines its representative with people from different disciplines or departments, rotating the leadership of the meeting every week. The many outside groups that observe the plant's Core Group meetings find them one of the best examples of participative management in action.

The growth of the plant to over 800 employees on three shifts has made face-to-face communication very difficult and recently prompted the inception of a daily in-plant newspaper. The paper, which disseminates information shared at the Core Group on issues such as deliveries, quality levels, and sales, is available about two hours after the Core Group meeting ends.

Other formal communication systems include regular manager, support group, and staff meetings.

One informal communication tool is the voluntary plant-wide meeting. At these monthly meetings, the plant manager discusses the plant's business performance and goals, answers employee questions, and periodically has customers demonstrate their products and discuss the needs of their business.

"Management by walking around" is another means by which Forest Grove managers encourage informal communication. Members of the plant manager's staff regularly visit the production and support areas, talking one on one with employees to better understand how the plant, from the technician's point of view, is performing. This accessibility enhances the team feeling at the plant.

Another informal communications tool is the widespread use of flipcharts. Centrally located within each Business Element, managers and technicians write notes on the flipcharts concerning equipment and process performance. This information alerts the next shift to specific production problems.

In addition, graphs depicting current production levels and plant manufacturing goals are strategically located throughout the plant. Other charts show the number of orders in each Business Element and their scheduled delivery date, enabling all Business Element members to be aware of late orders in their area. The result is that it is not uncommon for employees to volunteer to work overtime to rid their Business Element of backlog.

Individual and Group Development

Perhaps the biggest challenge any organization faces in implementing the principles of participative management is changing employees' attitudes. Key to changing the way people approached their work and performance were two distinct yet interrelated concepts. Bring resources together based on the work to be done rather than functional boundaries and the team concept, also known as self-managed work teams. These fundamental changes gave Forest Grove an unprecedented degree of flexibility in meeting the demands of its marketplace.

Structuring Resources Based on the Work, Not Functional Boundaries. Manufacturing performance at Forest Grove is measured by on-time delivery, product quality, contribution to profit, productivity, and quality of work life (Raynor, 1984). To achieve maximum performance in each of these areas, management developed a manufacturing approach that reinforced the interdependent relationship between employees, technology, and the current environment. Under this approach, the product is the central concern, and groups evolve under structures that facilitate the most efficient, highest quality production.

The new organizational structure divided the work by determining the point at which "a clearly defined and recognizable increment of value" (Hendrickson, 1982) was added to the product. The result was several distinct "Business Elements," each responsible for a different aspect of product manufacture.

This change also affected the relationship of such support groups as Engineering, Marketing, Human Resources, Training, and Quality to the production groups. Instead of functioning as autonomous groups as they did previously, each was integrated into the Business Elements.

The traditional structure in circuit board manufacturing and in much of the electronics industry is to divide work by function of job title. Tektronix' old circuit board facilities employed the "batch operation" approach, which was characterized by repetitious, assembly-line tasks. At the new Forest Grove plant, the "batch operation" system was eliminated and replaced with a continuous, process-oriented approach. The result has been improved productivity and higher quality of work life for employees.

The process-oriented approach requires a much higher degree of involvement on the part of its operators. In recognition of this, operators' responsibilities were expanded and their titles changed to "technicians." As technicians, they were expected to actively identify product defects at the earliest possible point, improve the processes, and be involved in their Business Element's decision making.

Expanding the roles and responsibilities of operators eliminated the need for several levels of managers whose primary responsibility was supervising their subordinate's work. About 150 inspectors were eliminated, as well as many extra staff. This flatter organization gave employees much greater control over their jobs.

The overriding objective of this approach is that every employee, whether in a production or support function, feels accountable for every product shipped to its customers, and that each shares equally in the success or failure of the entire plant.

Self-Managing Teams.   Under the old paradigm, production work was divided into simple, repetitive tasks performed by unskilled workers under close supervision. Managers made the decisions and directed the work; workers performed the work. Their access to information, ultimate responsibility for the work, and their knowledge and experience resulted in managers having virtual total control of the work situation. The major expectation of the workers was that they would take directions from the manager.

This paradigm was based on principles outlined by Frederick W. Taylor (1903). Taylor detailed his theory of "scientific management," an approach perhaps epitomized by Henry Ford's Model T assembly line. Ford's highly successful operation, based on his interpretation of Taylor's belief that workers should be considered simply extensions of machinery, greatly influenced American manufacturing practices for the next 70 years.

A changing environment and higher employee expectations are fundamentally changing this paradigm. In order to achieve or maintain its competitiveness and quality of life in the face of unprecedented competition from Japan, Europe, and even Third World producers, U.S. industry has been forced to re-examine its management practices. A workforce that increasingly demands interesting work and a voice in decision making is also driving the change in many companies from an authoritarian structure to an egalitarian environment that promotes accomplishment and trust.

While the "team concept" is not a new idea, in the experience of many workers its actual implementation varies with the whim of management, which tends to revert to the traditional hierarchical mode in times of crisis.

While the term evokes images of Japanese companies, the team concept is applied differently there than it is in the United States. In Japan, the major purpose of the team approach is to help detect problems in the production line to ensure quality. Another example of the Japanese application of the team concept would entail the members of a particular discipline, such as engineering, to work together as a team and cross train with the other members of their group.

Under U.S. participative management systems, however, team members have a voice in issues that in Japan are determined by management and engineers. Japanese companies are not driven from the bottom up, as was the Forest Grove approach.

At the same time that product flow through the plant was being restructured, the team concept was being introduced in each of the Business Elements and support groups. Under this new paradigm, the goal was for workers to become multi-skilled and manage themselves through teamwork.

In response to its customers' need for circuit boards which can meet their need for more sophisticated, powerful products, Forest Grove has been switching to a more advanced interconnect technology. In addition, the plant's products are increasingly customized to meet the unique requirements of each customer.

This changing environment requires flexible work practices and workers who are willing to learn all tasks within their team and to rotate from job to job. Members of each team are cross trained to perform all tasks, and can respond quickly to changes in production.

Cross training is implemented not only within each production area, or Business Element, but increasingly across disciplines. For example, employees who start out in, say, engineering or finance are encouraged to pursue opportunities in marketing, and vice versa. Consequently, the plant manager's staff, as well as managers plant-wide, have much greater understanding of all aspects of the business.

At Forest Grove, not only is the word "technician" replacing the word "operator," but "leaders" are replacing "managers." Leaders train technicians, share business information, and act as resource, coach, facilitator, and educator. Constant, effective communication is seen as the key aspect of leadership.

Technicians assume responsibility, participate in decision making, develop personal skills, and perform the work. To ensure that everyone has a voice in decisions that directly affect them, all decisions within the team structure are made by consensus. Involving employees in the decision-making process increases their contribution to the plant's daily performance.

A key element of Forest Grove's successful implementation of participative management has been its effort to keep in mind the personal stake for the individual. All employees are asked to play a major role in the cultural change to which Forest Grove is committed, change which centers on their willingness to learn beyond the relatively narrow confines of their own job and team to understand the entire business.

In this culture, all employees are held accountable for the success of the business in some way. This accountability is expressed in their understanding and actions, which are key to business success. According to Hendrickson, changing attitudes from a belief that employees are *entitled* (to pay, perks, etc.) to one that says they are *accountable* is the critical factor in affecting this cultural change.

In exchange for their efforts to understand the "big picture" and take responsibility for making an increasing contribution to the overall success of the plant, workers share in their benefits that accrue to a winning organization: high morale (for the organization as a whole) and higher self-esteem (for individuals); job security, good pay, and the right to be treated as professionals.

From its very beginning, the team concept at Forest Grove was not just applied to production and support groups, but to the management of the entire business as well. When the new plant was being built, a group called the Strategic Planning Group (SPG) was formed. Its major function was to design and implement strategies to win the support of employees for the proposed participative management practices, and to organize and implement operational training of employees.

In 1983, this group was replaced by the Resource/Process Planning Group (RPPG), whose primarily function over the next four years was to design an administrative infrastructure upon which the business could be managed. The RPPG was comprised of representatives from the Business Elements and support staff, as well as the plant manager and his staff, and became the core decision-making group.

Under Hendrickson's guidance, the RPPG developed the principles, philosophies, and strategies for the business. By its third year, the RPPG was responsible for the annual plant performance evaluation. (Certain functions of the former RPPG are still performed by ad hoc committees such as the Policy and Procedures, Awards and Recognition, and Compensation committees.)

As the needs of the plant changed, this central operational group evolved to meet those needs. (In the words of Gene Hendrickson, "Flexibility is holy, not the structure.") To reflect the plant's increasing customer focus, which was reflected in the plant's goal for fiscal year 1988—"To become every customer's best vendor"—a group called

STAT (Strategic Action Team) replaced the RPPG. The STAT's function included such responsibilities as implementing customer satisfaction teams to address customer issues and needs, initiating and facilitating strategic action plans and processes to increase customer satisfaction, maintaining a customer booth in the plant which provides employees with information about a different customer every two or three weeks, and overseeing the monthly plant meeting. All of these activities reflect the plant's increased customer orientation. Comprised of sales, marketing, and production people, the STAT group reported to the plant manager and his staff.

The STAT group disbanded in mid-1989, after it put the aforementioned activities and structure into motion. The activities have become a part of the culture and are supported by many individuals in the plant. The activities that focus on customer satisfaction are coordinated by the sales and marketing organization.

The fiscal year 1989 business goal is "to become established as the benchmark of excellence for the circuit board industry worldwide." Instead of having one central organization responsible for developing the strategies and tactics to reach this goal, each production and support group is required to formulate an action plan detailing how it plans to help attain this status.

Special Vendor Relations

The relationship between a company and its vendor is typically characterized by distrust and a belief that vendors are interested only in performing to a level which benefits them. This is not unlike the traditional relationship between management and workers.

Forest Grove believed that developing a positive, close relationship with a select group of vendors would have strong financial benefits. In exchange for its high expectation, often including customizing equipment and processes, Forest Grove promised to use single source vendors whenever possible. For vendors, this often meant they would be selling an entire integrated system instead of a single piece of machinery, which provided strong incentive for doing business differently.

One such example was the vendor training program. The majority of the process equipment introduced at the new plant in 1983 was completely new and unfamiliar. Under the vendor training program, vendors were on-site, overseeing the training of Forest Grove technicians, greatly reducing downtime by eliminating the need for a time-consuming internal training program.

A related activity was the certification program, which documented an operator's training and skill in effectively operating a given machine. Instead of a central training group developing and supervising the process, each Business Element designated a training subgroup responsible for developing and enforcing its group's certification program. Consistent with the goal of developing multi-skilled employees, all members of each Business Element are encouraged to undergo training and certification on multiple processes and/or machines.

Emphasis on Innovative Behavior

Under the old paradigm, management was considered to be not only all-powerful, but all-knowing. Under participative management, it is believed that the people who work most closely with the product have the best insight into ways of improving its quality and increasing overall productivity. At Forest Grove, the opportunity for technicians to act as innovators is facilitated by the fact they work with engineers rather than just taking instructions from them, as was the case in the past. One example of this cooperative working relationship is the increasing involvement of technicians at the early stages of the design process, ensuring that they work together with manufacturing engineers to improve processes.

Less tangible but no less important to nurturing innovation is the plant's non-threatening atmosphere. The communications channels and the team concept discussed earlier are critical to fostering an environment that does not discourage change and new ideas, but welcomes them. People are more likely to feel on an equal footing with co-workers in a non-hierarchical organization, and thus are more likely to voice the criticism and suggestions that lead to innovation. Risk takers are publicly acknowledged with certificates of appreciation and bonuses at periodic awards ceremonies.

Another method by which Forest Grove cultivates innovation is researching industry leaders in a particular process. Visiting competitors in order to improve their own processes has been widely practices by Japanese executives for years, but only recently have Americans begun to follow suit. At Forest Grove, not just managers, but engineers and technicians as well, travel to other U.S. and international companies to learn techniques and approaches that will improve the business.

Leadership and Support

Participative management systems are often inaccurately associated with weak, indecisive managers who have little authority over or impact on the decision-making process. In practice, the participative management approach requires strong, effective leaders who understand that by promoting an egalitarian, team-oriented atmosphere, they will reap increases in productivity and quality that will in turn reinforce their own position. Additionally, they will find that the scope of their responsibilities, as well as their personal skills, increases as their relationship to their team changes from controller/supervisor to facilitator/resource.

One of the most important roles of leaders is to develop and share their vision for the business and to motivate others to take the steps necessary to make it a reality.

Plant Manager Gene Hendrickson's vision for the plant led to this primary goal: "To become the world leader in circuit board manufacturing." In 1983, Forest Grove was so far from that goal that it was viewed as highly unrealistic by many inside and outside the plant.

Hendrickson recognized the need for what he called "champions," who would share his vision and help others understand and commit to it. As he developed processes, Hendrickson identified facilitators in each group charged with spreading the vision. They were trained in gathering, sharing, and presenting information and in leading the decision-making process. As workers began to understand their role in meeting the plant's goal, commitment and enthusiasm grew.

Not everyone was equally committed, however. As with any cultural change, there are people who don't alter their behavior to fit the new environment either because they don't grasp the significance of the change or because they choose to ignore it. In Hendrickson's experience, they eventually leave, having either been rendered incompetent in their jobs by their inability or refusal to adapt or having found the level of change in the environment too stressful to tolerate.

Hendrickson found that Forest Grove took two years longer than the three years he had anticipated to basically achieve this cultural transformation which, he stresses, is an ongoing process. He estimates that it takes a company of about 500 employees approximately five years to achieve "critical mass," and less time for a company with fewer employees.

The open, participative environment of Forest Grove has allowed leaders to emerge at all levels. By sharing visions and participating in the progress of its group, every individual in the plant is to varying extents realizing leadership potential, which is critical to the fulfillment of the plant's collective vision of a position of world leadership in circuit board manufacturing.

SUMMARY

Improved productivity, which is essential if U.S. industry is to achieve and maintain a position of strength in the world, as well as a high quality of life, requires investing in people, not just technology. According to Harvard University expert Richard E. Walton, "To have world-class quality and costs and the ability to assimilate new technology, we must have the world's best ability to develop human capabilities."

Participative management is based on developing and utilizing the capabilities of employees. The Tektronix Circuit Board plant has greatly improved productivity and quality and achieved a pre-eminent position in its industry worldwide through its commitment to the principles and practice of participative management.

This commitment is perhaps best stated in the plant's motto, imprinted on a large sign that greets every visitor to the plant: "What we're really doing here is developing people. Our business performance will be a measure of how well we do that."

REFERENCES

Belgard, William P., K. Kim Fisher, and Steven R. Rayner. 1988. "Vision, Opportunity, and Tenacity: Three Informal Processes that Influence Formal Transformation." In *Corporate Transformation,* Chapter 7, edited by R. H. Kilmann, and T. J. Covin. San Francisco: Jossey-Bass.

Hendrickson, G. 1982. "Business Guidelines." Memorandum (June 9).

Rayner, Steven R. 1984. *New Excellence: The Forest Grove Project.* Tektronix, Inc., p. 96.

Taylor, Frederick W. 1903. *Shop Management.*

Index